Small Farms

Rural Studies Series

Rural Public Services: International Comparisons, edited by Richard E. Lonsdale and György Enyedi

The Social Consequences and Challenges of New Agricultural Technologies, edited by Gigi M. Berardi and Charles C. Geisler

Rural Society in the U.S.: Issues for the 1980s, edited by Don A. Dillman and Daryl J. Hobbs

Technology and Social Change in Rural Areas: A Festschrift for Eugene A. Wilkening, edited by Gene F. Summers

Science, Agriculture, and the Politics of Research, Lawrence Busch and William B. Lacy

The Cooperative Extension Service: A National Assessment, Paul D. Warner and James A. Christenson

The Organization of Work in Rural and Urban Labor Markets, Patrick M. Horan and Charles M. Tolbert II

The Impact of Population Change on Business Activity in Rural America, Kenneth M. Johnson

Small Farms: Persistence with Legitimation, Alessandro Bonanno

Studies in the Transformation of U.S. Agriculture, edited by Eugene Havens with Gregory Hooks, Patrick H. Mooney, and Max J. Pfeffer

About the Book and Author

Why do small farms continue to coexist with ever-larger farming operations in advanced Western societies? Through a thorough case study of Italy and a comparative analysis of small farms in the United States, Dr. Bonanno seeks to answer this question, exploring the complex relationships among farm family members' ideology and behavior, the small farm economic sector, and the interaction of small farms within the relevant spheres of society. He finds that, at the structural level, a lack of occupational alternatives and contradictions within both labor and land markets often force farmers to retain marginal farms despite personal dissatisfaction. At the ideological level, many farm families display deep attachment to the agrarian way of life and cite this as a fundamental reason for not leaving the farm for other work.

Dr. Bonanno also analyzes the role of small farms within the social system and concludes that they serve a legitimative function. This legitimative role fosters contradictions within the social and economic systems that the state is unable to resolve, thus contributing to the continuation of a dual structure in agricultural development--large and very large farms at one end of the scale and marginal but persistent small farms at the other.

Dr. Alessandro Bonanno is assistant professor of rural sociology at the University of Missouri--Columbia. After undergraduate and graduate work at the University of Messina, Italy, he obtained master's and doctor's degrees in sociology at the University of Kentucky.

THE RURAL STUDIES SERIES of the Rural Sociological Society

Small Farms
Persistence with Legitimation

Alessandro Bonanno

Westview Press / Boulder and London

Rural Studies Series, Sponsored by the Rural Sociological Society

---

This Westview softcover edition is printed on acid-free paper and bound in softcovers that carry the highest rating of the National Association of State Textbook Administrators, in consultation with the Association of American Publishers and the Book Manufacturers' Institute.

Published in 1987 in the United States of America by Westview Press, Inc.; Frederick A. Praeger, Publisher; 5500 Central Avenue, Boulder, Colorado 80301

Library of Congress Catalog Card Number: 86-051502
ISBN: 0-8133-7341-7

Composition for this book was provided by the author.
This book was produced without formal editing by the publisher.

Printed and bound in the United States of America

The paper used in this publication meets the requirements of the American National Standard for Permanence of Paper for Printed Library Materials Z39.48-1984.

6 5 4 3 2 1

To the dedicated men and women
around the world
who live and work on small farms

Contents

Tables and Figures

TABLES

FIGURES

Acknowledgments

During the several years necessary for the research and writing of this book, I had the opportunity and privilege to count on the assistance of a number of friends and colleagues. I would like to acknowledge here those who directly or indirectly contributed to the realization of this project. Among them a special thanks goes to my teacher and friend, Larry Busch, who guided and advised me throughout the entire process. I am also indebted to Kathy Blee, Dwight Billings, David Dickens, Herb Reid and Lou Swanson, who read all or parts of my manuscript. Their comments were fundamental in eliminating a number of confused concepts, distorted understandings and misconceptions from the final version of the text. My Italian friends, Tonino Perna, Peppino Restifo and Amedeo Macri, spent several months with me in the field interviewing farmers. Their input made the interviewing process much more interesting and pleasant. I also would like to mention the contributions of the colleagues officially involved in the reviewing process for the Rural Studies Series. Their recommendations and remarks have been greatly appreciated. Writing a book, however, is not just carrying out a research project and conceptualizing the results; it also involves the long and arduous processes of typing and editing. On this front I was surrounded by a number of very competent and talented individuals. Ann Stockham carefully read a preliminary version of the book, eliminating stylistic imperfections and errors of distraction that are difficult to catch in a long manuscript. Pat Nelson and Debbie Garrett typed the entire manuscript and were patient enough to retype the revisions and addenda I kept putting on their desks. The most important thanks, however, goes to my wife Lucy Lee-Bonanno, who read all versions of the manuscript and carefully edited them. A great deal of her writing skills are embodied in this book. Thank you, Lucy, for your help. Finally, I would like to thank the farmers and their families who so openly and enthusiastically answered my questions. To them, to their commitment to farming, to their love for the land and their way of life this book is dedicated. As sole author I assume all responsibility for the content of this book.

Alessandro Bonanno
Columbia, Missouri

Introduction

This book examines the reasons for the persistence of small farms in marginal areas of one advanced Western country, Italy's Northeastern Sicily and Southern Calabria. Furthermore, it places emphasis on the international dimension of the phenomena, particularly with respect to the case of the United States.

Several considerations motivate this topic of study. First, emphasis has been placed on the decline in the number of small farms in most of the advanced Western countries (Fabiani, 1979; Fanfani, 1977; Klatzmann, 1978; Lin, Coffman, and Penn, 1980). This emphasis is not new in the literature, as students of the structure of agriculture both in the recent and less recent past have stressed a continuous decline in the number of small farms (Bogart, 1942; Ernele, 1936; Young, 1809). However, despite this trend it can be argued that the decline has not assumed a constant and gradual posture, for there have been periods in which the number of small farms has increased (Beale, 1978; Chantfort, 1982; Mottura and Pugliese, 1980). It can also be argued that the proportion of small farms within the structure of agriculture in most advanced Western countries has either remained virtually unchanged or, in some instances, has increased (Chantfort, 1982; Kautsky, 1971b; Mottura and Pugliese, 1980). Moreover, it has been stressed that in the future there will be a continuous expansion of the relative presence of small farms in the agricultural structure. In a discussion of the composition of American agriculture in the year 2000, a recent study concluded: "The farms will probably be arranged in a bimodal distribution-a large portion of small farms, an ever-increasing proportion of large farms, and a declining proportion of medium size farms" (Lin, Coffman, and Penn, 1980:iii). Consequently, despite an absolute decline in the number of small farms and despite the fact that their disappearance from the agricultural structure has been directly or indirectly implied for quite some time, small farms in advanced Western countries continue to represent a substantial and important portion of the agricultural structure.

Second, small farms are not homogeneously distributed within these countries (Bonanno, 1984; Klatzmann, 1978; Mottura and Pugliese; 1980). Instead, they are concentrated in marginal regions, namely, in geographical areas in which socio-economic indicators such as per capita income, gross local product (GLP), unemployment rate, and quality of life lag significantly behind the value of the same indicators for the country as a whole. It follows that the study of the persistence of small farms in these regions assumes a more relevant posture as they make up the primary portion of the agricultural sector and stand in several instances as one of the most important economic characteristics and sources of employment for the entire area (Calza-Bini, 1976; Mottura and Pugliese, 1975, 1980).

Finally, the continuous presence of small farms within the agricultural structure of advanced Western countries occurs despite the existence of agrarian policies in both Europe and the United States that discriminate against these units by assigning them the lowest amount of financial support among all farm categories. In the United States, for instance, "ten percent of the participants in Federal farm programs in 1978 received almost half of the farm program payments while 50 percent (those with small units) received 10 percent" (USDA, 1980). In Europe the agrarian policy of the Common Market has traditionally been oriented toward the development of large agricultural units and the elimination of small farms (Fabiani, 1979; Mottura and Pugliese, 1975). As a result of this strategy, financial support in the form of both price support programs and funds for the improvement of structures went primarily to large farms, leaving small farms, especially those in less developed areas like South Italy and Central France, with little or no support (Fabiani, 1979; Fanfani, 1977; Mottura and Pugliese, 1980; Zeller, 1970).

The task of analyzing the persistence of small farms cannot be separated from phenomena occurring inside and, above all, outside the farming sector. The importance of introducing a total approach to the study of this issue dictates the selection of dialectic as a theoretical paradigm of the book. Dialectic, is derived from the German idealistic tradition of Shelling and Hegel but here employed in accordance with Marx's later interpretation. For Marx, dialectic means the dynamic process through which reality is identified. It consists of three parts: thesis, antithesis, and synthesis, of which only the latter element is truly real. The synthesis gives meaning to the other two terms that constitute the position and the opposition, or negation, in the reality of the synthesis. This process follows an ideal, but not temporal, sequence of triads in which the latter triad completes and determines preceding ones and whose origins are the material conditions of production. Consequently, history for Marx is not generated by pure thought, as indicated by the Hegelian tradition, but is the history of human beings and their relationship with nature, a relationship determined by the action of people in pursuit of their interests. These interests are, however, antagonistic, so social classes favor and bear differing relationships with nature and human beings. History becomes the history of class struggle. The interpretation of social events is, thus, derived from the analysis of class struggle in a given historical context and from the relationship between structure (the economy) and superstructure (the ideological, political, and religious apparatus of a society) that specifies and is, in turn, specified by class struggle.

The dialectical approach differs from more traditional sociological approaches (the positivist-quantitative and the qualitative-interpretative) for two major reasons. First, it rejects the separation between object and subject in the process of investigation. A neutral posture in the observation of reality is not considered possible on the assumption that the knowledge of the observer, the observer him/herself, and the phenomenon in question form and are a part of the reality to be observed and, consequently, are not separable. Second, and related to the above, each scientific position is considered political, as it is ultimately generated by class interests. In the dynamic of class struggle, the dominant class becomes hegemonic, and its hegemony involves not only a domination in the productive sphere (control of the means of production) but also a domination at the superstructural level (Gramsci, 1973). It follows that science and knowledge are generated from the standpoint of the dominant class and are ultimately supportive of the interests of this class (Dickson, 1979;

Noble, 1977). As established by the tradition of the Frankfurt School (Adorno, 1973; Habermas, 1974; 1975; 1979; 1984, Horkheimer, 1972; Marcuse, 1968), the dialectical approach obtains knowledge through the critique of the false political neutrality of knowledge itself. In the dialectical approach man is not the product of the environment (objectivization) not the absolute legislator (Merlean-Ponty, 1973) (subjectivization) but a product-producer of a world that is historical.

These theoretical bases provide the framework within which analysis of the empirical data is carried out. Empirical evidence in the present work is derived from both observations arranged according to statistical procedures and from field notes arranged according to observational studies. Analysis of the data collected does not follow the procedures traditionally associated with statistical or observational studies. Rather, the dialectical approach calls for a data analysis that combines both methodologies, generating a completely new one. In this respect statistical arrangement of the data does not provide conclusive evidence but supportive evidence. Both this data and qualitative data are placed within a historical context, which acknowledges the dialectic of class struggle and the dialectical nature of the relationship between structure and superstructure. (Lukacs, 1971)

The opening chapter draws upon the major contributions of the topic that are present in the classical and recent literature. Though the issue of the persistence of small farms is not a new one, having been discussed at length by classic social scientists such as Kautsky (1971a; 1971b) and Weber (1958), the bulk of the debate can be found in works that date from the middle 1960's. This occurrence can be related to the crisis of development suffered by advanced Western societies during this period. After years of expansion of the forces of production, areas of underdevelopment, both at the structural and regional level, existed alongside developed areas. This phenomenon, which immediately became know as dualism, made apparent the inadequacy of theories hypothesizing homogeneous growth and attracted the attention of students in several disciplines. Concepts such as persistence and backwardness assumed a new and more relevant position within the sociological debate. In Italy this debate has been characterized by two principal positions. The first, which I will call ruralist (Barberis, 1970; 1974), assumes a homogeneity among the various categories of farms and, consequently, rejects the assumption of a qualitative difference between large and small farms. The persistence of small farms is, then, more the outcome of a voluntaristic act on the part of farm family members rather than a phenomenon related to economic or structural conditions present at the domestic and international levels.

The second position, which I will name conflictual (Calza-Bini, 1974; Daneo, 1969; Mottura, Pugliese, 1980) stresses the qualitative differences between large and small farms and the different roles that these two groups of farms play within the process of socio-economic growth. More specifically, for authors constituting this school, small farms tend to occupy a role that is related to the expansion or contraction of the economy. In periods of crisis and unemployment, the number of small farms tends to increase as they serve as a "container" for the excessive labor force. This process is also reinforced by the action of the State, which has to intervene in the economy by providing financial assistance to these farms. The persistence of small farms, then, becomes a phenomenon that is largely external to the voluntaristic sphere of farm family members and assumes a more structural and permanent posture.

The development and features of agricultural dualism in major advanced Western societies are the topics of the second chapter. In this context the qualitative differences between the small and large farm sector in countries such as the United States, West Germany, France Spain, and the entire Western European region are illustrated. The conclusions reached point toward the existence of an increasing gap between the two "faces" (dualism) of agriculture. One is that of the large farm sector, which controls the most conspicuous share of production, receives the largest portion of governmental aid, and tends to be concentrated in more fertile regions. The other face is that of small farms, whose contribution to production has decreased in past decades and whose share of State financial aid has been reduced. These farms tend to be concentrated in mountainous and hilly regions and areas with poor soil and constitute the largest portion of all farms.

These characteristics are particularly evident in the case of Italy and are illustrated in the third chapter. Here the phenomena of dualism, the decreasing productive capabilities of small farms, and the concentration of these economic units in marginal regions assume great importance in light of Italy's recurrent economic crises and uneven development.

The recent history of the Italian agricultural sector can be divided in three distinct periods. The first, which includes the years from the end of World War II to the end of the 1950's, is characterized by the reconstruction of the economic and productive apparatus after the destruction generated by the war. During this period the agricultural sector experienced rapid growth both in terms of total production and productivity. Small farms increased in number due to the implementation of agrarian reform, and new and more modern techniques of production were introduced. Small farms also provided jobs to a large portion of the rural labor force, though in numerous instances this employment took the form of underemployment. The major role performed by the small farm sector was, thus, that of "sponge" for the system; that is, keeper of an excess labor force now employed in a precarious manner.

The second period, which extends from the end of the 1950's to the middle of the 1960's, is characterized by a rapid economic expansion of all sectors, but especially of the industrial sector. This economic expansion, which was one of the strongest ever recorded among Western countries, was fueled by the availability of a cheap supply of labor within the country. Consequently, a substantial portion of the labor force one kept underemployed in agriculture was released and transformed into an urban proletariat. The small farm sector thus lost its role as "sponge" for the system and underwent a process of rapid transformation. By the end of the 1960's, the total number of small farms was reduced by over a million units from the figure of a decade earlier, the land farmed was reduced by 3.5 million hectares, and the total contribution to the GNP by small farms drastically shrank.

The third period, which includes the years from the middle sixties to the present, is characterized by the development and consolidation of a dualistic agricultural structure. In these years the bulk of agricultural production was increasingly concentrated among large farms, whose number diminished in the South and increased in the affluent North and whose output accounted for the vast majority of agricultural revenues. The small farm sector occupied a marginal position both economically and geographically as its productive role shrank. Small farms became increasingly concentrated in poor areas, particularly in the South. During this period small farms once again assumed the role of keeper of excess labor who could not be employed in other economic sectors due to the serious economic crisis experienced by the country since the early seventies.

The methodological posture of the book is discussed in the fourth chapter where specific details regarding the empirical portion of the work are introduced. In this context the characteristics of the region surveyed are illustrated, as well as the techniques of data collection and analysis. Sampling procedures and the results of the statistical elaboration of the data are also presented.

The fifth chapter contains representative cases of farm families interviewed. These families are divided into six groups on the basis of their motivation for remaining in farming. The first group, which is named "Traditional," is composed of families who keep their farm due to emotional attachment to the land and/or farm life. The second group, called "Reserve," contains families who do not have available job alternatives to farming or do not have skills that will allow them to leave the sector. The third group, "Complementary," consists of families who remain in farming because the income from this activity complements that of family members with off-farm jobs. Families classified in the fourth group, which is referred to as "Retired," are characterized by the fact that one or more members have retired from a previous occupation, and they have invested savings in the purchase of a farm or they have a farm they already own but have not actively cultivated. The fifth group, that of "Equity," is composed of families who have invested in the land, though their reasons for doing so differ from those of the previous group. The sixth and final group is a "Residual" group. In consists of several subgroups of families who remain in agriculture for differing reasons. Among them, three assume particular theoretical relevance and are discussed in detail. The first is the subgroup of families who view farming as a hobby to be pursued during leisure time. The second consists of families who are ready to sell their farms and will do so at the first available and convenient opportunity. Finally, there are families who would like to sell their farms but consider the offers received not adequate to the real value of their farms.

The chapter concludes with a critical discussion of the evidence presented. It suggests that the conclusions reached by classic and modern sociologists who dealt with the phenomenon of persistence of small farms are largely applicable to the situation of South Italy and Sicily, though only a synthesis of these theories can fully account for the phenomenon in question.

The sixth chapter discusses the persistence of small farms from the point of view of their position within the social system. Through the use of the concepts of accumulation and legitimation, it is emphasized that a very large portion of small farms do not accumulate; rather, they perform a legitimative role in society. The performance by small farms of the role of legitimation fosters a number of contradictions. Despite the intervention of the State, these contradictions cannot be removed from the system without creating new ones in different societal spheres. It is argued, then, that the State is not able to balance the system, nor can such a balance be reached automatically by the system itself (ineffectiveness of the so-called invisible hand). Consequently, some kind of disequilibrium must be maintained within the system. As far as agriculture is concerned, the existence of disequilibria is translated into the continuous existence of a dualistic pattern of agricultural development that, in turn, implies the continuing persistence of small farms.

The seventh and final chapter examines the implications and perspectives that the analysis developed for Italy has in regard to the case of the United States. Through a further development of the discussion on the role of the State in advanced societies, a micro and macro analysis of the persistence of American small farms in undertaken. Finally, the relationship between farming and other socio-economic sectors is explored, emphasizing the process of industrial decentralization and informalization. The importance of the notion of growth with underdevelopment is stressed as one of the most important conditions for the reproduction of the present pattern of expansion of the forces of production.

1
The Persistence of Small Farms and the Debate in the Literature

THE PHENOMENON OF PERSISTENCE

In the advanced societies of the Western world, there has been a historical trend toward a decrease in the number of farms and, among them, especially of small farms (Lin, Coffman, and Penn, 1980; Mottura and Pugliese, 1975; Pfeffer, 1983; Rodefeld et al., 1978). Moreover, the structure of agriculture in these societies has been characterized by a relative growth of large farms, which became dominant in the sector, and by the marginalization of small farms, which have decreased in number at a very substantial rate (Buttel and Newby, 1980; Chantfort, 1982; Crecing 1979; Fabiani, 1979; GAO, 1978). The explanation generally used to account for such trends emphasizes the economic weakness of small farms and their inability to cope with the increasing competition of larger farms, which is associated with the concentration and centralization of capital that shapes advanced industrial societies. However, the marginalization of small farms and their expulsion from the sector do not imply their total elimination. The shape of several, if not all, of the advanced agricultures in the world tends to assume a dualistic form with a pole of large farms and an opposite pole composed of small farms.

The observation of such phenomena is not new. Kautsky, investigating the trend of European agriculture at the turn of the century, wrote: "I have reached the conclusion that we should not expect, in agriculture, the end of the small farm nor the end of the big farm. At one pole, we have the universally true tendency toward proletarianization, but, at the other, we have a continuous swinging in the development of small and big farms" (Kautsky, 1971a:9). Kautsky's observations were intended to counter the prediction of European sociologists and economists who, basing their analysis on the assumption of the rigidity of the laws of competition, forecasted the total disappearance of small farms as a result of the growth of larger and more efficient units. Weber also argued against the disappearance of small farms. In a discussion of the changes taking place in the structure of agriculture at the time, he identified the small farmer's quest for personal freedom from the "patriarchal relationship of personal dependence" as the factor responsible for his endurance of extreme living conditions in order to maintain independence (see Giddens, 1971).

Since the writings of Kautsky and Weber, the mechanistic tendency to picture the development of agriculture as a linear trend has been increasingly regarded as unsatisfactory. The alternative interpretation that has emerged sees agriculture as a sector within the overall socio-economic process of development, which is related to the other socio-economic sectors of the of the system in such a way that the development of one of them is affected by and influences the development of the others (Buttel and Newby, 1980; Friedlard, Furnari, and Pugliese, 1980; Mingione, 1981). In this respect agriculture does not follow the development observed for the industrial sector as if the sectors were two independent units following the same pattern. Instead it is engaged in a trend parallel to the development of industry and yet different in the sense that the two modes of development are faces of the same process (Mottura and Pugliese, 1975).

Within this framework, the persistence of small farms is not to be seen as a somehow temporary situation serving as a transitional point from a stage of general predominance of small units to a stage of exclusive presence of large and/or corporate enterprises. Rather, it is to be seen as a permanent situation in which the persistence of small farms is a product of the development of advanced capitalist societies, and small farms, therefore, play an important role in the process of reproduction of the system (Fabiani, 1979). The persistence of small farms is, in our historical period, an end in itself rather than a means to an end.

THE PERSISTENCE OF SMALL FARMS IN THE CLASSIC LITERATURE

In the classical sociological tradition, the notion of persistence holds a somewhat marginal position. Nineteenth century theories of social change, in fact, were strongly based on the notion of evolution, which defines change as a general process of growth toward more complex and sophisticated forms of society (Hoogvelt, 1982; Portes, 1976). These theories implied a unilinear development, that is, one line of growth that recurs in every society. They equated change with progress so that any societal change would be automatically positive and beneficial and would create the conditions in which a higher level of civilization could be attained. Comte's three stages of development (1970), Spencer's evolution from homogeneous to heterogeneous (1967), and Durkheim's (1964) change from a society based on mechanical solidarity to one based on organic solidarity are the most well-known examples.

Within this framework, the persistence of a societal unit was seen as a leftover from a previous stage of development, an element of backwardness eventually destined to disappear in the future when the process of transition to the next stage would be fully completed. Moreover, as the pattern of development was the same for all societies, a delay in change was attributed to the lack of specific elements that were characteristic of more advanced societies. The focus on attention was, thus, placed on the inevitable progress and on the "modern elements" characterizing it rather than on the traditional ones. Traditional elements that persisted, assumed either a romantic posture, as in the "Gemeinshaft" of Toennies (1957), or a negative one as in the positive philosophy of Comte (1970). They occupy a central position in neither, however, for their centrality was in contradiction with the postulates of these theories and with the assumptions of the general philosophy behind them. Evolutionism was also a main element of theories of change developed within the Marxist framework. The economic determinism of the late Engels and that of the Second International are cases in point.

As far as the farming sector was concerned, evolutionary theories stressed the inevitable and desirable growth of large farms, which implied the consequent disappearance of small farms. During the nineteenth century the Western world experienced very rapid economic growth that involved the expansion of industrial activities and the establishment of large-sized urban areas. Faced with the problem of an increasing population, the major goal at the time was for all the European countries (the United States represented a case apart)[1] to produce cheap and abundant food, given the lack of sufficient land. The solution to this problem provided by most social scientists was summarized in the words of Arthur Young, a British student of agriculture: "...large farms, large capital, long leases ad the most improved methods of cultivation and stockbreeding" (Young, 1809:175). His theory, which rapidly became very popular, stressed that the goal to be achieved was that of developing to the "utmost the resources of the soil" and that social considerations should be subordinated to it (Ernele, 1936).

The popularity of this doctrine was also justified by the constant growth of large farms and the decrease of small farms that took place in Europe almost without interruption from the middle 1700s to the 1880s (Bogart, 1945; Levy 1911). In this context small farms and small farmers were seen as impediments to the growth of agriculture, thus, as obstacles to be overcome. In the words of a contemporary of the period, small farmers were: "...slow to discern signs of the time, and to modify their old routine in accordance with the new circumstances" (Brodrick, 1881:378).

The Marxist position on the topic is shown in the Erfurt Program. The Erfurt Program was the document with which the German Social Democracy established itself in 1891 as the leader of the working class movement all over Europe (Brusco, 1979). According to the Program, all workers who were small commodity producers (petty bourgeoisie) in both the industrial urban sector in agriculture were destined to become wage workers. Large industrial enterprises and large capitalist farms would lead small enterprises inevitably to bankruptcy and their owners to destruction. "This is the inevitable outcome of the economic development of modern society, as inevitable as death" (Kautsky, 1971a:53).

In the mechanistic assumptions of economic development that inspired the Program, the creation of socialism as the new stage of social civilization would be reached automatically due to the inner contradictions of the capitalist mode of production. The role of the working class and the socialist party was to remove all obstacles to this process and to facilitate its rapid conclusion. Accordingly, these principles opposed a delay in the proletarianization of the petty bourgeoisie and the process of concentration and centralization of capital, which would precipitate the fall of capitalism. Small farmers, as well as urban small producers, were not to be included in any social or political program sponsored by the socialist movement (Procacci, 1971).

Paradoxically, the position of socialists against small farms was stronger than that of other political formations which clearly did not belong to the left. In fact, in several intellectual circles close to the European political establishment of the time, the need to have a larger small farm sector was expressed (Levy, 1911). This concern arose from the problem created by the massive rural out-migration that occurred during the period and the consequent congestion of urban areas (Bogart, 1942; Dobb, 1947). Especially after the economic crisis of the 1880s, which affected all the major countries of Europe, the object of delaying the flux of rural immigrants to towns became particularly important. Consequently, several legislative measures were taken in favor of small holdings. In England the Allotments Act was introduced in 1887, the Small Holding Act followed in 1892, and the Local Government Act was passed in 1894. All of these were directed toward the creation of small holdings. Programs to facilitate credit for farms with limited capital were also introduced (Levy, 1911). In Germany the government created the Land Mortgage Association (Landshaften), which "served the interests of peasant cultivators" (Bogart, 1942:284). In France legislation favored the creation of cooperative organizations and credit associations from which small farmers benefitted greatly. In all of these countries, as well as in the United States, educational programs for farmers were developed. It must be stressed, however, that these programs were mostly instrumental in maintaining social control. The opposition they received from the socialists was due, therefore, to the basic socialist goal of aggravating the contradictions present in society in order to destabilize it.

The contributions mentioned so far have one thing in common. They do not contemplate the possibility of the persistence of small farms but only consider the advantages or disadvantages of the presence of large farms and small farms. It is only at the end of the nineteenth century that the notion of persistence of small farms began to appear in the social debate. An initial reason for this event was that, in spite of economic expansion and recurrent crises, the disappearance of small farms was slower than that of other small urban artisan enterprises and never fully occurred in any country (Brusco, 1979:172; Guaraldo, 1979:162). Second, and in connection with the above, the lack of empirical support of the linearity of development opened the way for attempts to reformulate the theories of growth abandoning the predominant evolutionary posture.

This phenomenon first occurred within the Marxist school, where new ways of proposing the relationship between small and/or impoverished farmers and the working class were tackled. It is within this framework that Kautsky, the main author of the Erfurt Program, revised his position in The Agrarian Question (Die Agrarfrage), which was originally published in 1899. Kautsky made clear that the analysis of the Erfurt Program lacked an understanding of the social situation in agriculture and did not take into account the differences that characterized such a sector (Kautsky, 1971:16). From these premises he reached the conclusion that small farms and large farms would be constant characteristics of capitalist agriculture.

The explanation for such a conclusion is related to the role that capitalism assigns small farms, that of producing a wage labor force to be used by large capitalist farms. Small farms, thus, persist because: "The owners or tenants of small farms are in the best conditions to have several children and educate them to rural work" (1971b:185). He also stressed the continuous need that large farms have for labor produced by small farms and explicitly criticized programs generated in those years in favor of small holdings. "Wherever large farms eliminate small farms in great numbers, conservative politicians and rich landowners undertake actions, both privately and publicly, to increase the number of small farms" (1971b:188). Kautsky concluded that the condition of autonomous independent workers (petty bourgeoisie), which small farmers have is only a judicial facade covering their real position of proletarians. On the basis of this new analysis Kautsky asked for a revision of the previous political strategy of the German Social Democratic Party against small farmers. Such a revision was made and the strategy changed.

Non-Marxist contributions supporting the persistence of small farms also appeared. Levy published in 1904 the German edition of Large and Small Holdings which was followed in 1911 by the English edition. Levy's analysis was aimed at countering both theories that predicted the complete destruction of small farms by large farms and theories that supported the superiority of small farms over other types of farms. While acknowledging that the system of large farms had made continuous progress ever since the eighteenth century, he also stressed that from the 1880s onward these conditions were altered. Using England as a case study, he pointed out that as a consequence of the

European economic crisis of the 1880s and 1890s, the number of large farms and very large farms decreased, while the number of small farms increased.

The occurrence of this phenomenon provided him with evidence that Young's theory was false. As a supporting explanation for his position, he analyzed the different productive advantages of the two types of farms and correlated these advantages with the historical conditions present at the time. In his analysis changes in the structure of agriculture were due to "the changes in the condition of sale and production since 1879" (Levy, 1911:154). Since that time there was a decrease in the profitability of dairy, fruit, and vegetable production. The latter, as he suggested, "prove to belong, on the whole, to the domain of small farmers, who on the other hand only undertake corn growing as an altogether secondary matter" (Levy, 1911:155).

Given these conditions, a number of landowners broke up their properties and converted once large farms into small farms, while owners of small farms were able to prosper. As a conclusive remark Levy noted that the conditions dictated by the "market" would influence the growth of one group of farms over the other but that one group would never have total supremacy.

Paradoxically, Levy's analysis was, in a sense, complementary to that of Kautsky. While the former dwelled on the "economic reproduction" necessary for the persistence of small farms, the latter emphasized "social reproduction" as a condition for the persistence of such farms. Both points of view, however, would be used in a revised form in the future as bases for contemporary analyses of the persistence of small farms.

In spite of these works and others that addressed the issue in a less direct and articulated manner,[2] the concept of persistence of small farms was rarely used in the literature until the early 1960s. There was, of course, a lengthy treatment of small farms, their role, and their trends within agriculture in the Western countries. All of these analyses focused, however, on issues such as the desirability or disadvantages of the existence of a small farm sector, the legislation and incentives to create more small holdings, the measures needed to improve the productivity of small farms, their social advantages, etc.

It was only in the sixties that the debate on the persistence of small farms was revitalized. The reasons for such a revitalization are somewhat similar to those that prompted Kautsky and Levy to undertake their works. They can be summarized by the fact that after years of expansion of the forces of production, areas of underdevelopment, both at the structural and regional level, existed alongside developed areas. The phenomenon, which immediately became very well known under the term of dualism, attracted even more attention because it existed not only in the third world but also in advanced industrialized countries. Within the member countries of the European Community, for instance, several underdeveloped areas were officially identified in the course of the last two decades and numerous regional plans of development were designed and executed without providing an overall solution to the problem (Clout, 1976).

In the United States similar patterns were also followed; the case of the Appalachian region serves to illustrate them. In these areas rural communities and farming represented, and still represent, an important portion of the total economic activity. This circumstance suggested numerous studies of the relevance of agriculture and small farms. Italy is among those countries in which the existence of development and underdevelopment has had strong repercussions on the overall process of growth within the country and in which small farms have played a key role. The recent debate on the persistence of small farms in Italy is the topic of the next section.

THE CURRENT DEBATE IN ITALY

In the past two decades the debate in Italy on the persistence of small farms has been characterized by two major approaches. The first, which will be labeled ruralist approach assumes a posture that identifies the farm sector as homogenous with a continuum of small-large farms. The second approach, which will be called conflictural, identifies a contradiction of interests between the social groups connected with large farms and those connected with small farms.

The ruralist approach is very well represented by the works of Barberis (1970, 1974), who described the structure of Italian agriculture by using census data (1974). He classified farms according to two criteria: farm tenure and participation of the farm operator in any manual farm activity. The picture that emerges from this classification is that of a very homogeneous farm sector in which over 80 percent of all farms belong to the category of family farms (aziende contadine). The other 20 percent are classified as capitalist farms, where farm manual activities are executed by hired workers. No reference is made to the amount of manual work done on the farm by the operator, to the incidence of hired work over the total amount of farm work, or to the total sales and size of the farm. Barberis's goal was to demonstrate that in Italy there was a trend toward "the expansion of petty commodity producers formed by small farm owner-operators" (1974:21), and that "we are in presence of...agricultural artisans" (1974:17).

In this context the structural distinction between large and small farms is denied. The existence of prosperous farms and of marginal farms is explained only in terms of ability to manage the farm. Accordingly, the persistence of small farms is a technical problem related to the will and, above all, to the ability of the farm family (Bertolini and Meloni, 1979). Farm mobility (i.e., the possibility of expanding or reducing the economic size of the farm) is left to the discretion of the farm operator. Barberis stated: "The two realities (large and small farms) are a continuum without interruption: the small family farm does tolerate a certain degree of capitalist[3] management: the large capitalist farm does not exclude family work" (1974:17).

The contribution of the ruralist approach did not satisfy the Italian audience completely. To many it appeared evident that a better understanding of the "agrarian question" could not be achieved without an evaluation of the differences among social groups and among farms. This awareness formed the bases for the development of the conflictural approach. Within this approach, however, are orientations that, while keeping a common denominator, differ substantially. A review of some of the major contributions of this composite approach follows.

A position that bridges the ruralist approach and the conflictual approach is that provided by Brusco. His Rich Agriculture and Social Classes, which was published in 1979, contains a specific section on the persistence of small farms. It should be noted, however, that the topic of his research is the farm structure of advanced regions and, consequently, the conclusions reached may be not pertinent to the object of this study, the persistence of small farms in marginal areas. Nevertheless, the overall relevance of the book makes the reference worthwhile.

Brusco stressed the idea that the persistence of small farms can be explained as a consequence of the action of the State (legislative, political, and legitimative measures) in their behalf. This phenomenon, which was frequent in several advanced Western countries in the very recent past, boosted the number of small farms and in some cases improved the conditions of life of the farm family. However, as Brusco continued, it is more difficult to explain the persistence of small farms when farm policies are not specifically addressed to small holdings but involve all types of farms in a general fashion. He noted that the latter trend has been dominant in all the advanced countries in the last two decades. Consequently, the aforementioned explanation lacks empirical support. Moreover, arguing from a more economic point of view, he countered the position that sees governmental farm support programs as always somehow beneficial to small farms.

This position argues that price support programs and other similar programs generate high profits for large farms but also a sufficient income to small farms. Brusco suggested that this hypothesis is true in the short run but becomes inadequate if we consider a longer period of time. In fact, in spite of the lack of fluidity of the land market, "Over a sufficiently long period of time, the high profits of large farms will attract new operators in the large farm sector" (Brusco, 1979:164). Consequently, the large farm sector will expand, while small farms will diminish rapidly. Thus, the explanation for the persistence of small farms must be found elsewhere.

His conclusion was that the phenomenon of persistence is due to the desire of farmers to work longer hours in order to obtain additional income. The small farmer has a subjective understanding of what is a "just" remuneration for his work. If this remuneration is higher than the one actually obtained from the farm, he will abandon farming. Conversely, if it is equal or lower, he will remain. This explanation, Brusco continued, would justify the persistence of small farms even in conditions of low remuneration of labor per hour and low productivity. It is not important to the farmer that his work has a lower remuneration per unit of time than that of other hired workers provided his total remuneration is superior to that of hired workers. In this way small farmers enjoy the possibility of expanding their income, if they chose to do so, with additional work.

In order for this explanation to be effective, however. several conditions need to be met. First, workers' attitudes toward working hours must be such that some workers are willing to work a relatively limited number of hours and receive a wage for that amount of hours, while others are willing to work more hours for a higher remuneration. Second, all farmers must have the possibility of finding off-farm work in farming or other jobs outside the sector if they wish to do so. Finally, wages in farming must equal those of other sectors. Brusco emphasized that these conditions are met only in limited cases. Nevertheless, they are present in the area he investigated (the Modena region in Northeast Italy).

Brusco's theory contains a full recognition of the differences between types of farms and a clear awareness of the different implications that farm policies have, depending upon the type of farm involved. Nevertheless, his explanation of the persistence of small farms tends to emphasize the individualistic and voluntaristic theme typical of the ruralist approach. His account, which in many respects recalls that of Weber, makes persistence a decision totally in the hands of the operator. Structural elements are, once again, reduced to secondary features. Moreover, his economic approach oversimplifies the picture as a whole, and its social aspects, other than those involved in the analysis of the psychological decisionmaking process of the farmer, are omitted. However, it should be pointed out in Brusco's defense that his analysis may very well summarize some of the characteristics of Persistence in "rich" areas where a dynamic labor market and possibilities of off-farm employment are in plentiful supply.

Another important contribution is that of Daneo. in 1969 Daneo published Agriculture and Capitalist Development in Italy in which a clear and effective break with the ruralist approach was made for the first time. Daneo's work addresses three main issues connected with the structure of agriculture. The first involves a discussion of the composition of the small farm sector vis-a-vis that of the large farm (capitalist) sector. He identified in Italy the existence of a large group of small farms that covers, in his estimate, 42 percent of all farms. This figure is substantially smaller than that provided by ruralist analyses (80 percent) using the same census data. The make-up of the capitalist sector is estimated at 38 to 40 percent of all farms, while an intermediate stratum, identified by Daneo as peasant-captialists, occupies between 18 and 20 percent of all farms. He stressed, in opposition to the assumptions made by the ruralists, that this groups is the one internationally known as family farms and that all other small farms, while generally run by a farm family need to be treated differently.

The second issue discussed by Daneo concerns the formation of a dualistic structure within Italian agriculture. He states "The majority of farms are concentrated on two opposite poles, that of small farms and that of large farms" (1969:148). He also suggested that a structural dualism as marked as that of Italy is not found in other advanced societies where medium-sized farms make up a large portion of the total farm sector.

Finally, Daneo addressed the issue of the persistence of small farms. In a large number of units of the small farm sector, there is an excess of labor (i.e., hidden unemployment). People in this group are marginalized by an income of a mere subsistence level often supplemented by off-farm employment. In these conditions the farm is more a statistical datum than an actual economic unit (1969:18). However, the subsistence or even growth of the farm is guaranteed, according to Daneo, by the intervention of the State. The State, through the welfare system and special programs, finances these farms and the hidden unemployment that exists there. This action is taken "...in the name of the common interests of all capitalists to maintain a political and social equilibrium in society" (1969:158). Accordingly, in mature capitalism there are productive units that are "...economically destined to defeat but are socially condemned to reproduction" (1969:24).

Daneo's analysis introduced new elements into the debate on the persistence of small farms: the action of legitimation conducted by the State and, implicitly, the issue of the role of the State in mature capitalist societies. The action of legitimation, which is defined as the guarantee for the reproduction of the conditions of social harmony necessary for accumulation (O'Connor, 1973), can be conducted by different spheres of society (Poulantzas, 1978). However, the State has always had a major role. Daneo singled out the interaction between the State and the farm sector (understood as the social fractions involved in it), which is aimed at reducing the contradictions present in society and establishing a pattern of development that favors specific fractions of the bourgeoisie over the working class and petty bourgeoisie.

Daneo's work influenced several other authors, among them Bolaffi and Variotti who published their Capitalism Agriculture and Social Classes in Italy in the early seventies. The book presents a case study of the social stratification within Italian agriculture, using as its subject a rural region near Rome. The authors, following Daneo's suggestions, identified three different structural sectors within agriculture: large, small and medium (peasant-capitalist) farms. However, they made a further distinction within the small farm sector in their identification of a group of very small farms, which they call "improper" (improprie), a group of part-time farms, and small farms per se.

There are two criteria upon which their differentiations are based. One of these is the absolute quantity of labour used on the farm, while the other is the ratio of wage labor over total labor used. Furthermore, in their analysis small farms per se have a clear disadvantage with respect to large farms. This situation is due to several factors. First, there is a difference between the two types of farms related to the advantages that economies of scale can provide to large farms vis-a-vis the use of capital and labor. Second, large farms have more flexibility in hiring the required amount of labor for each specific season or farm task, while small farms always have a constant supply of labor. Finally, the inferiority of small farms over large farms is established by the inability of the former to introduce technical innovations at the same rate as the latter. However, in spite of the various disadvantages that small farms display, their presence in the structure of agriculture has not changed. Bolaffi and Varotti, in fact, stressed that from 1948 to 1970 the relative size of the small farm sector remained essentially the same. This observation enabled them to conclude that the process of concentration predicted by orthodox Marxists never occurred in the agriculture of the region they studied.

Conclusions by Bolaffi and Varotti (1973:211) point out two major characteristics of modern farming, the precarious and subordinate character of small farms on one hand and their persistence on the other. For them the existence of small farms is not a transitional phenomenon but a permanent one (1973:233). As far as the reasons for persistence are concerned, Bolaffi and Varotti rejected the explanation provided by Daneo when they stated that the persistence of small farms "...cannot be explained in terms of political mediations which tend to preserve the socio-political equilibrium of the system through the support of unproductive sectors" (1973:233). Their alternative explanation was, however, not free of problems. They stated that small farms persist because the prices of agricultural products "are a function of the productive stratification of the sector as a whole, namely, that the price somehow expresses a mean of the costs of production" (1973:205). This explanation is very effectively criticized by Brusco, who comments

> To say that the price is formed around the mean of the costs of production of farms does not explain the reason for the survival of small farms. It is not clear why, in light of the possibility of gaining higher profits, large farms do not out-compete small farms and eliminate them from the market. In the new situation, the price of agricultural products would still be determined by the mean value of the costs of production, but this time the producers would be only large farms, which have a higher level of efficiency.

Other aspects of Bolaffi and Varotti's theory are ripe for criticism, as well. Some of the reasons for the subordination of small farms are rather questionable. The advantage that large farms have over small farms with regard to the flexibility of the use of labor is one of them. Studies of both Italy and other Western countries support the view that family labor typical of small farms is actually an advantage rather than a disadvantage. In a study of Northeastern American farms, Colman and Elbert concluded: "Family labor remains ideally suited to the sequential farm production cycle, because family members can and do sustain periods of intense labor followed by periods of enforced layoff" (1984:5). Similarly, their allegation that the introduction of technological innovations is always in favor of large farms is not entirely convincing. Small farms can have, in certain cases, a technological edge that cannot be achieved by large farms. (see Freidland, Furnari and Pugliese, 1980:23).

Bolaffi and Varotti's analysis represents a clear rejection of Kautsky's suggestions. A "Kautskian approach" is adopted, however, by Calza-Bini (1974, 1976), who is the author of several important contributions on small farms that appeared during the middle seventies. Calza-Bini rejected the explanation that small farms persist because of technical characteristics that enable them to compete successfully against large farms. Instead, he supported the idea that the persistence of small farms should be explained by analyzing the conditions of the labor market and by connecting the trends present there with the general pattern of development of mature capitalist societies.

In the first of the two works considered here Calza-Bini (1974) observed that "Small farmers remain in farming because they do not have other job alternatives" (1974:93). He developed this conclusion in his other work (1976) by stressing the theory, first developed by Mottura and Pugliese (1975), that agriculture as a whole performs a double function for the system. The first function is that of serving as producers of agricultural commodities and, second, as keepers of the surplus labor force.[4] It is because of this second function, he continued, that we can explain the contraction, expansion, and persistence of the small farm sector that occurred in the agrarian history of Italy. However, in order to understand persistence fully one must to take into account the phenomenon of industrial decentralization, which takes place in advanced capitalist societies. The process of concentration of productive units has been reversed in past years so that tasks once concentrated in one unit are now decentralized and delegated to a number of others in both rural and urban areas. This process creates, among other things, a "marginal" labor market (part-timers, underemployed), which is constituted largely by a farm labor force. Accordingly, Calza-Bini state: "The phenomenon of persistence of small farms occurs due to ... the possibility of farm family labor making use of the marginal labor market" (1976:28). Moreover, the persistence of small farms and productive decentralization are complementary phenomena. In fact, Calza-Bini stressed that if decentralization generates the persistence of small farms, the abundance of marginal labor available in the latter creates the conditions for the existence of decentralization. He further stated: "the progressive socialization of industrial work, takes advantage of the socialized characteristics of farm work ... and makes farming a possible option" (1976:47). If the process of decentralization continues, there will be a consolidation of the dualistic structure of agriculture. The productive aspect of farming will be carried on more and more by large farms, while small farms will provide an underemployed labor force available for alternative occupations. This trend not only guarantees the persistence of small farms but it will generate a further expansion of part-time farming, which will involve a considerable number of ex full-time small farmers.

Calza-Bini also observed that small farmers are actually proletarized and that their position of marginalized independent workers is just a facade that covers a different reality. The process of proletarianization of farm family labor takes place, according to the author, in three stages. The first consists of a reduction of the farm income, which lowers the standard of living of the farm family. The second stage is represented by "production in loss" in which the farm is still kept as an economic unit but does not produce profits for the family. Finally, the farm is transformed into a subsistence farm in which its products are used exclusively for self consumption by the family. At this point farm labor is released into the market, though only partially, and keeps fluctuating from farm underemployment to precarious industrial jobs in the decentralized sector.

Calza-Bini's analysis is very well articulated and brings a new light onto the debate. It is extremely important not to isolate processes taking place in the farming sector from others occurring in other socio-economic spheres, both domestically and internationally. His connection of change with persistence and the identification of the new characteristics of such a relationship is an important contribution. On the negative side, however, his analysis gives the impression that small farms always perform the role of keeper of surplus labor than just in some cases. This position, which at this point strikingly recalls that of Kautsky, forces the analysis to the point that the typical diversification of the small farm sector suddenly disappears, leaving in its place a new and unexplained homogeneity.

Probably, the most intelligent and articulated analysis of the persistence of small farms and the structure of agriculture in Italy is that of Mottura and Pugliese. In a collaboration that spans almost two decades, the two authors have produced a significant number of contributions that have been translated into several languages. Here, their book Agriculture, South Italy and the Labor Market, is considered. It was originally published in 1975 and contains several essays that earlier appeared as separate contributions. An English summary of their work is also available in Buttel and Newby's The Rural Sociology of the Advanced Societies (1980).

The Mottura and Pugliese position was derived from Marx, particuarly Chapter 23 of Capital, and from Gramsci. Their introductory goal was to reject as false the two extreme positions developed in both the classic and the modern literature. The first of these argued that the fundamental law of capitalist development (concentration and centralization of the means of production and proletarianization of labor) takes place in agriculture in the same way as in industry, while the second supported the idea that the future of agriculture belong not to large but to small farms (1975:13).

Their assumption was to relate dialectically the development of the forces of production in all socio-economic sectors with the historical conditions determined in society and then to identify the trends in the farm sector. Given this assumption, they agreed with the conclusions reached by Kautsky on the persistence of small farms but departed from the argument of the German social scientist when stating the reasons for the existence of such phenomenon. They argued that in capitalist agriculture there are two major roles historically performed by the farm sector as a whole. The first role is the productive one, which involves the production of agricultural commodities. When this role is predominant in society, there is an expansion of large farms that are productively superior to small farms. Small farms, in these periods, are considered as structures that guarantee the survival of a portion of farm labor that does not have the possibility of finding a more remunerative job in other sectors in the short run. The second role is that of keeper of the industrial reserve army. In periods in which this role is predominant, there is an increase in the relative volume of small farms, mostly provided by the introduction of programs in their favor which also generate an improvement in the conditions of life of small farm families. The contrast with Kautsky's position now appears very clear. Small farms do not generate a surplus labor force that is destined to be used by the large capitalist farms. Rather, they keep a labor force that, if release, could create further contradictions in society.

However, Mottura and Pugliese did not limit their analysis to only these considerations. They tried to relate the variations within the small farm sector to the general process of economic and social development. First, they noted that the persistence of small farms does not deny the general trend of proletarianization of the rural petty bourgeoisie but makes this phenomenon appear in a different form than the one suggested by orthodox Marxists. Accordingly, in some cases of part-time or subsistence farming, there is a situation in which small farmers keep their apparent condition of petty commodity producers, but are in reality proletarianized. Second, persistence does not imply a superiority of small farms over large farms or an ability of the former to out-compete the latter. There is in capitalism a subordination of the small farm sector to the large farm sector. This situation is supported by the action of the State and other socio-political structures, both domestically and internationally. Mottura and Pugliese have illustrated this last point by analyzing the action of the European Common Market vis-a-vis agrarian policies. The

Mansholt Plan, as well as the "Directives" that followed its application, were in their view aimed at the expansion of large farms and at the reduction of the number and overall volume of small farms.

Finally, in spite of the differences within the farm sector, there is a general subordination of agriculture to industry, which is generated by the faster circulation of capital occurring in the secondary sector. This dependency, according to the authors, creates a different degree of power at the class level among agricultural and industrial fractions of the ruling class. Industrial groups tend to be hegemonic over farming groups and influence the country's policies of development toward their interests, thus, relegating agriculture to a somewhat secondary role.

Mottura and Pugliese have been so influential in Italy in the last few years that several authors have followed their steps and further developed their conclusions. Among these, the works of Gorgoni, Furnari and Mingione deserve mention.

Gorgoni (1977) emphasized the importance of separating the different fractions that form the small farm sector. Part-time farmers, marginalized farmers, and the remainder of small farmers are sometimes contrasting realities, behind which lie different motivations for their presence in farming. Moreover, Gorgoni stressed the distortions that the agrarian policy of the Common Market creates within the farming system. He concluded that price support programs and "orientation programs" of the EEC contribute in a contradictory and partial way to the persistence of small farms.

Furnari (1977) focused her attention on the conditions of the labor market and underlined the difficulties that farmers have in finding alternative jobs outside agriculture. She also pointed out that there are regional differences. The labor market is less fluid in the South where the overall economic conditions are worse than in other regions of Italy. There, the persistence of small farms is largely related to the lack of job alternatives and the depressed economy.

Mingione (1981) reviewed the types of farms that exist in Southern Italian agriculture and emphasized that the rural areas of the region are not homogeneous. This phenomenon creates a variety of motives for the persistence of small farms. In more dynamic areas, which are usually flat and coastally located with good infrastructure, small farms can survive due to the economic ability to make a somewhat sufficient profit. In less dynamic areas the survival of small farms is guaranteed by price support programs and, above all, by welfare payments, which in some cases make up over 70 percent of the net income of farm families. In these conditions the farm provides resources (such as food for home consumption or a limited profit from market exchanges) that, complemented with welfare payments, create the possibility of persistence. The situation is then considered the best possible choice for farmers because of the lack of job alternatives.

The debate on the persistence of small farms is not limited to Italy, but is as developed and informed in many countries as that described above. In the United States, for instance, in the last decade a number of publications and research projects have tackled directly or indirectly this issue (for a discussion of the case of the United States, see Chapter 7). A large body of literature which examines the phenomenon of persistence through the analysis of part-time farming (Cavazzani, 1980; Coughenour and Gabbard, 1977; Fuguitt, et al, 1977; Schroeder, et al, 1985) has pointed out that the persistence of small farms is connected to cultural, emotional and ideological postures of farm family members. According to Coughenour (1977), for many small scale farmers the economic rewards attached to the farm operation are of secondary importance. Such a conclusion is paralleled by Crecink, who in a discussion of the income of small scale farmers indicates the low importance of the economic dimension of farming for small farm families. Farming as a lifestyle emerges as a central factor for the persistence of small farms in the work of Schroeder, Fligel and Van Es (Schroeder, et al, 1985). In their view economic factors such as the maximization of profit have a marginal bearing on the question of persistence, given the limited ability of farmers to optimize their resources and obtain sufficient income for the farm household. These and other studies (Smith and Capstick, 1976; Swanson and Bonanno, 1986; USDA, 1985) also indicate that non-profit maximizing strategies have been adopted by a substantial number of small scale farmers.

Structural factors, however, have also been employed in several cases to explain the persistence of small farms. Paralleling Mottura and Pugliese's explanation for the case of Italy, Beale (1978) indicates that during the great depression the persistence of small farms was related to the role of keeper of surplus labor that they performed. Agricultural economists (Mapp, et al, 1979; Raup, 1978) have pointed out that small farms can actually survive on the market, as their organization of production can be optimized better than that of larger farms. Similarly, small farmers can, in some instances, introduce innovations and adapt to the change conditions of the production process faster than larger producers, thus gaining a surviving edge (Friedland, et al 1980).

The literature review presented in this chapter examines the complex and often divergent positions held by authors concerning the persistence of small farms. However, a few common points emerge, the most important of which is dualism. Almost all of the authors who support the persistence of small farms see the structure of agriculture as polarized into two large aggregates; small farms and large farms. The centrality of dualism is also established by the shared conviction that the persistence of small farms and the continuous existence of large farms is not a transitional phenomenon but rather a permanent one. Chapter 2 focuses on these considerations by analyzing the characteristics and meanings of dualism in the farm sector in some advanced Western countries.

2
Dualism in Some Advanced Western European Countries and in the United States

The concept of dualism employed in this book is used specifically in reference to its structural and territorial dimensions. Structural dualism indicates the existences on one pole of the farm spectrum of a limited, yet expanding, large farm sector which produces a considerable portion of the total agricultural sales and/or occupies a substantial area of cultivated land. It also implies the existence on the other pole, of a decreasing, but large, small farm sector. Structural dualism is further characterized by a collapsing middle farm sector. The territorial dimension of dualism refers to the contrast between developed and underdeveloped regions within a single national territory.

In order to further assess the importance of dualism in agriculture, the conceptual and empirical problems related to the concept of small farm are explored in this chapter. Furthermore, a brief description of the agricultural structure of selected advanced western societies is undertaken. It is the object of this final portion of the chapter to demonstrate the significance of dualism in today's agriculture in the western world.

THE NOTION OF SMALL FARM

In the attempt to describe and analyze dualism in the agricultural structure of advanced Western societies, it is relevant to address two important issues. The first involves the identification of criteria for the definition of small farms, while the second deals with the position of small farms within the socio-economic system. In the survey conducted in South Italy for the present study, small farms were identified as farm households with an annual product sold in 1982 of less that 20 million lira and with a land area of more than 1 hectare. The definition was developed in consideration of the specific situation of Italian agriculture and the criteria traditionally used within the literature on the topic. However, this definition is only one among several used in recent analyses. In fact, while there is a general common understanding of what a small farm is, the notion assumes different meanings when used within the context of a particular research project. Daneo (1969), as well as Mottura and Pugliese (1975), defined small farms according to the acreage of the farm. Bolaffi and Varotti (1973) used two criteria. One was the absolute quantity of labor used on the farm, while the other was the ratio of wage labor over total labor. Fabiani and Gorgoni (1973) employed three different measures of the quantity of labor. Finally, the Italian Census Bureau (ISTAT) has published data that define small farms according to their tenure and acreage but not according to their total annual sales.

It has been noted (Mingione, 1981) that all of the criteria used for these definitions present problems. When the acreage of the farm is used, there is a tendency to underestimate the differences in output and revenue that exist among farms of the same physical size but with different internal organizations and cultural practices (10 hectares cultivated in grain do not equal 10 hectares of fruit of vegetables or flowers) (Klatzmann, 1978). When the criterion is the amount of labor (total and/or hired labor) used on the farm, there is a tendency to overestimate the number of small farms. This phenomenon is due to the development of technological devices that reduce the amount of work required on the farm so that hired labor and/or part of family labor are not utilized.

The problem of selecting criteria for the definition of small farms becomes more difficult when an international analysis is attempted. The various linguistic, cultural, and, above all, historical differences from country to country generate concepts of "small" and concepts of "farms" that may describe very different entities. This conclusion is not new. Bogart (1942) commented on these terms in a table on the structure of agriculture in France at the turn of the century: "An American, or even a British, reader are at once struck by the classification of this table, which lists holding of 100 acres as 'large'" (1942:292). Moreover, he added: "These differences are due to historical... factors, but they also show the difficulty of making international comparisons and of drawing too hasty conclusions" (1942:292). In Europe, however, the tendency to classify farms according to their physical size has prevailed over the decades.

In spite of the theoretical and methodological objections aimed against this practice, the major bodies of data issued domestically and internationally on the structure of agriculture retain physical size as the key criterion for distinguishing between small and large farms. This attitude toward data classification has been reinforced by the strong positive correlation that exists between size and farm income (Fabiani, 1979; Russi, 1971). In the United States small farms have been defined using various criteria and identified as family farms, families with limited resources, farms with a limited volume of agricultural sales and farms with a limited acreage (Carlin and Crecink, 1979; Munoz, 1984). However, the most common criterion used for the definition of small farms is their total annual sales (Heffernan et al., 1982).

How is it possible, then, to compare European farms with American farms given the diversity of criteria used in the collection and publication of data? How can a satisfactory common definition of small farms be achieved when the statistics available in Europe classify farms according to their size only? One solution to the problem would be to use the common available data, which allow a definition of farms based on their physical size. Such an approach, however, would generate a classification in which most of the European farms would be counted as small farms (90 percent of all European farms are under 100 acres), while most of the American farms would be considered large (only 12 percent of American farms are under 25 acres). More importantly, this classification would eliminate totally all of the cultural and historical characteristics that differentiate European and American farms. These characteristics are, in fact, at the root of both the composition of agricultural structure and the manners in which such a structure has been analyzed.

The alternative approach which we would like to adopt reintroduces historical and cultural elements in the criteria for farm definition. In this analysis of dualism, farms are classified in accordance with the general criteria used within each country so the European small farms are identified on the basis of physical size, while American farms are classified in relation to their total annual sales. There are, of course, problems related to the use of these criteria, but, given the data available, they provide the most suitable solution.

The second issue is that of the position of small farms within the socio-economic system. In recent years two major approaches have emerged suggesting different solutions. The first approach, which derives from the analyses of the "Dependency" school (Frank, 1979; Wallerstein, 1974), stresses that capitalism has been the dominant mode of production worldwide since the eighteenth century and, consequently, all elements present in society should be considered as capitalist. Small farms, in spite of their productive organization, which resembles a non-capitalist form (petty commodity producers), are capitalist. There are, therefore, no relevant differences between large farms and small farms vis-a-vis their organization of production. Both types of agricultural units are based on some form of private property. Hence they are capitalist (Sherry, 1976). This position has received a degree of criticism that is primarily addressed to the rigidity with which the entire socio-productive structure was analyzed. The most recurrent of these is aimed at the inability of the dependency approach to acknowledge the historical co-existence of capitalist and pre-capitalist forms of production within the same society. The social formations that originated from such a co-existence are, then, key elements in understanding the domestic and international patterns of development in all socio-economic sectors (Taylor, 1979).

The second approach to the position of small farms in the system considers small farms as an element differentiated from capitalist production. This approach, which has received considerable attention among North American social scientists (Goss, Rodefeld, and Buttel, 1979; Rodefeld et al., 1978), can be summarized in the workds of Friedmann:

> Capitalist production involves two classes, one which owns the means of production and another which labor; the two are connected through the wage relations, in which an entrepreneur purchases labor power from others in order to set in motion his means of production. Household production involves only one class, which both owns the means of production and provides labor power to set them in motion; relations of production with the

> enterprise are based not on the wage contract, but on kinship. When the household is specialized and competititive, and means of production and subsistence must be purchased, it is simple commodity production (1978:548).

In other words, according to this latter position a structure is considered as capitalist only when it displays the typical "capitalist" characteristics, namely concentration of the means of production and wage labor. Moreover, this approach reaches two relevant conclusions about the role of small farms in society. The first involves the notion that the presence of small farms represents an impediment to the expansion of capitalism and, consequently, to the propagation of the contradictions associated with this mode of production (Nikolitch, 1972). The second conclusion argues that small farms are the perfect type of economic unit capable of restoring a pure form of democracy according to the Jeffersonian ideal (De Janvry, 1980).

The posture taken in the present work rejects both positions while retaining some of the assumptions characteristic of each of them. The opinion presented here is that small farms are an integral part of the capitalist mode of production and they do not represent a world apart from the rest of society. They do have important characteristics, however, that differentiate them from other productive units.

In order to support this position, it is worthwhile to reporpose some basic concepts and conclusions presented by Kautsky in his Agrarian Question (1971). In this work Kautsky stressed that what is generally understood as agriculture today can be defined as the set of operations of soil cultivation and animal breeding and that the products derived from such operations are called "agricultural products." He noticed, however, that this definition does not define agriculture in general. Rather, it is a historical defintion limited to the capitalist mode of production. In fact, it is only with the rise and establishment of capitalism that a gradual diversification of agricultural activities from industrial ones occurred.

This phenomenon was accomplished through the separation of peasant handicrafts from urban manufacturing and commerce, and it was subsequent to this separation that agricultural products became commodities. This phenomenon occurred because farmers needed to buy an increasing quantity of manufactured goods and to pay taxes to a State that no longer accepted payments in "kind" (Mottura and Pugliese, 1980). This development increased the farmer's need for money, and simultaneously, his need to sell products in order to get it. In fact, given the existence of industry, the products that a peasant could sell easily were those that came from the cultivation of soil and breeding and not the ones that came from his activity as a craftsman. This situation was due to the fact that industry could produce the latter more easily and cheaply. So, as Kautsky stated, the farmer had to become what is understood today as a farmer, a person who grows crops and raises animals.

Given these assumptions, today's farms, and consequently small farms, are part of the capitalist mode of production neither because they display the "typical" characteristics of that mode of production not because they are based on private property but because their present form was generated by the development of capitalism and because they exist within capitalism.

This conclusion avoids the rigid formulation typical of the dependency approach by reintroducing the possibility of the existence of non-capitalist productive relations within capitalism. The social formation that derives from such a co-existence of different relations of production has capitalism as the dominant (hegemonic) mode and pre-capitalist and/or post-capitalist relations as subordinate ones. By the same token, the existence of small farms within capitalism supports the idea that small farms do not represent a world apart from the rest of society but, rather, display some specific characteristics in their existence and development that differentiate them, as well as the farming sector as a whole, from other productive units in other socio-economic sectors. Such a difference is due to the specific character of agricultural production, which is linked to biological cycles that serve as a brake to the velocity of circulation of capital, from which the rate of profit originates (Man and Dickinson, 1980; Mottura and Pugliese, 1980).

In non-farming sectors the rate of profit can be expanded by increasing the number of complete cycles of production, and, indeed, this is one of the major goals of any entrepreneur in the process of accumulation. In farming the ability to reduce production time is restricted by the biological nature of such production, which can be only partially overcome through the introduction of technical innovations (Mann and Dickinson, 1980; Mottura and Pugliese, 1980). Furthermore, as Mann and Dickinson pointed out: "Some spheres of agricultural production are also characterized by a significant gap between total production time and labor time so that for many agricultural commodities there are lengthy periods where the application of labor is almost completely suspended..."(1980:286). This situation leads to the intermittent employment of labor and machinery and consequently, under-utilization of capital and lower levels of accumulation than those that could be realized in other sectors. Thus, "It is not surprising that those spheres of agricultural production marked by these features tend to be left in the hands of petty producers" (Mann and Dickinson, 1980:287). The integration of the small farm sector within the capitalist system and the specific characteristics of small farms form, then, the elements from which the analysis of dualism in agriculture should be initiated. It is, in fact, through the evaluation of the process of development in capitalism and the specification of the ramifications of the overall process of dualism that the present features of the structure of agriculture in Western countries can be understood.

REGIONAL AND STRUCTURAL DUALISM

The concept of dualism has generally been used to refer to uneven levels of development, first at the regional level and later at the structural level. However, its meaning differs according to the paradigm within which it is applied. In the modernization approach dualism indicates the lack of contact between advanced and backward regions in which the latter do not experience growth due to their traditional traits that differ from those of the former (Allen, 1974). The solution to such an unbalanced situation lies in increasing contacts and diffusion from the developed regions (Pye and Verba, 1965). As far as the structure of agriculture is concerned, the modernization approach calls for the elimination of dualism through both the elimination of backward farms that cannot meet the modern standards of production because of their size or location and the consolidation and technological improvement of others that display potential for development (Barberis, 1970).

According to the dependency approach the differences in development that characterize dualism arise at the same time and are generated by the same process. There is, in other words, a simultaneous and intertwined development of some regions and underdevelopment of others. Frank has summarized this positions:

> Underdevelopment is not due to the survival of archaic institutions and the existence of capital shortage in regions that have remained isolated from the stream of world history. On the contrary, underdevelopment was and is still generated by the very same historical process which generated development: the development of capitalism itself...These regions did not have a chance against the forces of expanding and developing capitalism, and their own development had to be sacrificed to that of others (1972:8 and 12).

In farming the co-existence of areas of advanced agriculture, mainly characterized by rich large farms, and of those of backward agriculture, characterized by poor, small farms and the large-small farms dichotomy itself, is due to the overall process of development that channeled resources from less advanced areas and structure to more advanced ones (Mingione, 1981).

The concept of dualism used in the present work stems from that of the dependency theory yet departs from it is some respects. First, the relationship between development and underdevelopment is not viewed in such a mechanistic fashion as is typical of the dependency approach. Development and underdevelopment, rather than being causally related, are dialectically related (Dickens and Bonanno, 1983). That is, the occurrence of development generates underdevelopment, but at the same time the occurrence of the latter affects the specific forms in which development takes place. Second, the dependency approach insists on economic factors (structure) as the only elements generating dualism. Here, ideological, cultural, and political factors (superstructure) are considered to play an important, and in some instances determinant, role in the formation of development and dualism (Gramsci, 1974). Finally, dualism is seen as generated, in the last instance, by the struggle between classes (and/or their fractions), rather than being the outcome of the law of capitalist development as is often suggested by dependency studies (Amin, 1978; Frank, 1979).

Given the epistemological context indicated above, the next step in the analysis of structural dualism in advanced Western societies is that of stressing the relationship between regional dualism and structural dualism in agriculture and the related qualitative differences between the large and the small farm sector. Regional dualism is a rather diffuse phenomenon among Western countries. In Western Europe, in spite of the consistent and unprecedented development that occurred in several countries during the last three decades, there has been a growing concern with the differences within and between regions (Clout, 1976). Among the member countries of the European Community, even a simple examination of population density shows a fundamental core/periphery contrast in which heavy urbanized regions exist together with sparsely populated regions such as Central France, Scotland, and almost the entire territory of the Republic of Ireland (Denton, 1969).

From a financial point of view, regional dualism is evident to a still greater extent. Denmark is the most affluent country with a GNP about 28 percent above the EEC average, while West Germany (+23 percent), France and Luxemburg (+18 percent), and Belgium (+8 percent) all report GNPs above average. Conversely, Great Britain, Italy, the Republic of Ireland, and Greece fell below that mean (C.E.C., 1979). Almost all of the countries in the EEC contain regions that display Gross Local Product (GLP) values above and below the Community average. The richest region is the area of Greater Paris, while portions of Denmark, West Germany, Great Britain, and Italy show high GLP levels. South Italy, Central and West France, Scotland, and Ireland score well below average. Similar patterns are found in the United States as well, for the per capita income and the GLP vary greatly from the low levels of the Southeast to the high levels of the West (Ross, Blustone and Hines, 1978).

Coupled with underdevelopment in all of the advanced Western countries is the presence of poor, small farms. Clout has summarized the phenomenon:

> Some regions are dynamic, with growing cities and flourishing industries, but others are not so 'healthy', being characterized by high unemployment, low average wages, loss of jobs in old established industries such as farming and coal mining, and important outmigration, as people move to dynamic areas and try to improve their living standards (1976:1).

This passage specifically refers to Europe, but the same conclusions can be reached for the United States. Counties where 80 percent or more of the farm operators run farms with annual sales under $20,000 tend to be located in regions that have a lower than average per capita income (USDA, 1978). Traditionally underdeveloped areas in the US like the Appalachian region display similar characteristics with very poor, small farms, out-migration, and coal mining. Farming as such does not mean underdevelopment, but small farming is associated with underdevelpment. Rich farms are characteristics of a number of advanced regions both in Western Europe and in the US. The dairy farms of Denmark and West Germany, the vineyards of South France and North-Central Italy are all cases in point.

The dualism of small-large farms is not, then just a structural differentiation between two groups of farms, but it assumes a specific connotation in the overall process of socio-economic growth. It implies that small farms are qualitatively different from large farms. This difference is displayed, first of all, by the fact that small farms tend to be located in regions in which alternatives and potentials for development are generally scarce (Franklin, 1971). Second, social inequalities exist between developed and less developed regions and between small and large farmers. Finally, each farm group faces different roles and problems.

The relative importance of agriculture in advanced Western societies has diminished, however, in the past decade both in terms of population employed in the sector over the total employed population and in terms of total contribution to the GNP. Nevertheless, it represents by far the most extensive rural use of land and a fundamental economic sector in both Western Europe and the US (Clout, 1975). In Western Europe 44 percent of the total area is devoted to agricultural production, and this percentage exceeds 70 percent in countries such as Spain and Ireland. Moreover, in spite of a reduction in the population employed in agriculture throughout the region, agricultural employment still represents a substantial portion of total employment in some countries. France, Italy, and Spain alone contain three-fifths of Western Europe's farmers and farmworkers, who represent sizable political interests in Western European affairs (Franklin, 1971).

Within each country regional inequalities create a situation in some areas in which the portion of workers dependent upon jobs in agricultural activities is well above the national average. Such are the cases of Western France and South Italy where the percentage of people employed in agriculture is almost double that of the whole country (Fabiani, 1979; Klatzmann, 1978). In the United States the percentage of the population employed in agriculture is substantially lower than that of all the Western European countries with the exception of Great Britain (Table 1). This situation arose from the historically extensive character of American agriculture (Klatzmann, 1978; Tracy, 1964) and from the different pattern of socio-economic development that occurred in the US, which favored extensive mechanization and a more rapid substitution of men with machines than that recorded in Europe (Goss and Rodefeld, 1977: Mann and Dickinson, 1980). However, US agriculture is by far the most productive of all the Western countries, as it is the only one in which the percentage of population employed in agriculture does not exceed the percentage of agricultural production figuring in the GNP (Table 1). Furthermore, the percentage of people directly employed in agriculture conceals the fact that agriculture in the US in the last several decades has been an important part of both US foreign and domestic policies and that currently "Agriculture is the largest industry in the country employing some 17 to 20 million American workers directly or indirectly" (GAO, 1978:4). Finally, the socio-economic importance of agriculture is not equally distributed throughout the territory, making the "agrarian question" in the US a regional as well as national issue.

Agricultural structural dualism in advanced Western countries signifies the existence, on one hand, of a limited yet expanding large farm sector which produces a considerable portion of the total agricultural sales and/or occupies a substantial area of cultivated land. On the other, it represents the presence of a decreasing, but large, small farms sector (Tables 2 and 3).[5] The importance of a dualistic agricultural structure, as well as its origin and effects of the overall societal organization, varies from country to country. On one pole there are countries such as Italy, Spain, or West Germany in which dualism is undoubtedly one of the most important characteristics of the local agricultural structure. In other countries such as Great Britain there is a totally different structural arrangement. Great Britain represents, however, an atypical case in the agrarian configuration of advanced Western societies. It was the first agriculture in Europe to complete the transition from feudalism to capitalism, which occurred in the early nineteenth century considerably before other Western European countries (Ernele, 1936; Tracy, 1964). Moreover, the industrial development that characterized the British economy at the time and the resulting rise in the demand for food made agriculture a profitable occupation. Substantial capital was invested in agriculture which in a relatively short period of time was transformed into a sector characterized by "large unified farms, many of them in possession of wealthy landlords who were prepared to invest heavily in improvements to their estates" (Tracy, 1964:42).

The situation changed briefly during the great depression of 1880-1900 when, due to international competition (especially from the United States and other extra-European countries such as Australia and Argentina) the prices of several agricultural commodities, above all grain, dropped sharply. Large farms decreased in number, while the number of small farms increased quite substantially (Levy, 1911). The reaction of all of the European countries to the crises was to adopt protectionist policies through the introduction of import tariffs. This phenomenon, however, did not occur in Great Britain. In fact, the British government implemented a free trade policy that within only a few decades generated a substantial reduction of the agricultural sector to the point that it became the smallest among all of the advanced Western countries. The reasons behind such a political decision are largely related to the particular strength of the British urban bourgeoisie over the rural classes as a whole and to the fact that the overwhelming majority of the working class was, at the time, an urban-industrial working class (Clapham, 1932). For the urban bourgeoisie, as well as for the urban working class, "cheap bread" was a war-cry, and serious pressure was applied against the repeal of the Corn Law or any other protectionist act. Faced with this competition, both owners of large holdings and small farmers did not have political power to influence the government's decision (Tracy, 1964). The free trade policy was subsequently abandoned, however, and the protectionist policies that followed were neither strong nor changed the structural composition of British agriculture, which consisted primarily of large farms.

In the same period, the various agrarian policies adopted by the other European countries were influenced by the rural classes' large political, economic, and social power. Heavy demonstrations were carried on by farmers and farmworkers all over Europe, and pressure was exerted by landowners on the various governments to take protectionist measures (Tracy, 1964). Protectionism has since that time become a constant characteristic of most Western countries and guarantees, among other things, the persistence of farms that if left to face open competition from abroad, would disappear in a short period of time (Birnie, 1961).

The existence of protectionist policies explains only partially the differences in the structural composition of agriculture between Great Britain and other advanced Western countries. The formation of a dualistic structure in these countries is related to the specific characteristics of the historical process of development that took place in each of them. In very general terms, it can be said that the growth of a large farm sector has been fostered by the fact that these farms provide levels of accumulation far superior to those that can be achieved by other types of farms. Accumulation is also the major goal of capitalist society and of the capitalist State (Holloway and Picciotto, 1977; Offe and Ronge, 1979; Poulantzas, 1978). Hence, agrarian policies in Western countries have been in one way of another supportive of the large farm sector (Fabiani, 1979; GAO, 1978; Mottura and Pugliese, 1980; USDA, 1980).

On the other hand, the presence of a sizable small farm sector is related to social legitimation. Small farms in all the Western countries represent either an ideal form of socio-economic activity (the Jeffersonian ideal, Agrarianism) and/or an element that guarantees a certain degree of political and social stability for the system in situations in which alternatives are lacking (Bonanno and Ritter, 1983; Mingione, 1981a; Mottura and Pugliese, 1980). A classic historical example of social legitimation is provided by the events of 1848-1949 in France:

> French peasants were in favor of the Second Empire (a new hegemonic strategy established by the industrial bourgeoisie against the former financial establishment), because they remembered the advantages that they had gained during the First Empire. But the economic and social situation had changed so much that they could not get from Napoleon III any reinforcement of their weak economic position. During the First Empire the small farmers' productive structure could be implemented by the expropriation of large landlords, by war and war economy. Fifty years later, small farmers were being expropriated (proletarized) and weakened by international competition in food production through imports from America and Asia, independently of any government policy. In this sense the Second Empire regime was established on a hegemonic

project combining the interests of expanding industrial capital with only the purported interests, but active support, of the farmers, who, at the time, constituted the large majority of the French population (Mingione, 1981a:41-42).

French small farmers' support of the Second Empire was based on the conviction that this regime was in their favor. The farmers' ideology (in this case expressed by their political beliefs) legitimated the rule of Napoleon III, one of the most reactionary regimes in Europe in the last century. Another and perhaps more familiar case of social legitimation is that of small farms in the United States during the Great Depression of the 1930s. During this period there was an increase in rural population. This growth "seems to have occurred because conditions in the cities were so dire that the subsistence possibilities in the rural areas looked comparatively attractive. Outmigration was deferred until better times, and some return flow from the cities was received" (Beale, 1978:39). During these years the small farm sector provided subsistence possibilities to a large portion of the population who would not otherwise have been able to survive. Migration to rural areas helped to minimize the contradictions in the system and to create a minimum level of stability. This form of legitimation was instrumental in avoiding an increase in social tension, while creating conditions that allowed for future economic growth.

A more recent example of social legitimation that involved small farmers is the case of "peasantization" in Italy. On two different occasions in the recent past, the Italian ruling class used the support of peasants to legitimize its power. During the 1930s the Fascist regime obtained the support of the rural population (most of whom were landless peasants working on latifundia) by assigning them small lots of land. These lots were not sufficient to guarantee any economic future to the peasants but were enough to make them feel as though they were independent farmers. Their historical goal of owning land had been achieved. This made them grateful to the regime that allowed their dream to be fulfilled. The rural masses consistently supported fascism until the regime collapsed during World War II. Similarly, in the second post-war period the Christian Democratic Party (DC) obtained a large political consensus from the rural masses by means of a land reform, which provided small parcels of property to landless peasants. This agrarian reform was not successful as far as redistribution of land was concerned (see Chapter 3). However, it was able to establish a pro-DC sentiment among the rural population in spite of the fact that the policy of this party was aimed at protecting other social strata.

The specific forms in which structural dualism in agriculture appears today and the qualitative differences between small and large farms is presented in the next section, using the cases of some selected advanced Western countries.

DUALISM IN WESTERN EUROPE AND IN THE USA

The countries surveyed are among the world's most advanced. There are, however, differences among them that need to be taken into account with regard to both their economic development and position in the international economic arena. The differences determine the relative importance of agriculture within the national context and the importance of the small farm sector.

The United States

The structure of American agriculture can be summarized as follows. Approximately 80 percent of all farms in the United States can be considered small farms. They control nearly 50 percent of the farmland resources but receive less than half that percentage in total value of farm sales (USDA, 1978). However, 6.6 percent of the nation's farms control nearly 27 percent of the land and make up nearly 54 percent of the total market receipts (USDA, 1978). Furthermore, the average farm income for the largest farms is nearly 600 times greater than that of the smallest farms. These differences between small and large farms have increased tremendously over the past 20 years. It is evident, then, that the structure of American agriculture tends to be "characterized by two major classes of farms: large farms, producing most of the nation's outputs, and small, part time farms..." (Chantfort, 1982:4).

The most recent census data show that in relative terms, while mid-size farms diminished as a result of the recent economic crisis, small and large farms increased their presence within the sector (Chantfort, 1982). Large farms have been expanding during the past few decades and recently, the very large farms, those with a total annual sales of more than $500,000, have displayed a substantial increase. In 1969, only 0.1 percent of all farms were members of this last category, while in 1978 they represented more than 1 percent of US farms, increasing from 4,000 to 25,000 (USDA, 1978). The numerical increase is not as large, however, as their increase in the share of farm production. While these farms control 8 percent of US farmland, they make up 30 percent of farm market receipts, up 16 percent from the 1969 figure. Furthermore, they make up 60 percent of the national farm net income, up 44 percent from a decade ago (Velde, 1983). The largest farms also tend to be geographically concentrated with about 20 percent of them located in California, 7 percent in Texas, and an additional 33 percent in only nine other states (Velde, 1983).

Small farms are also growing. This growth, however, is qualitatively different from that of large farms. In fact, it appears to be related to the capacity of survival of some farms and, above all, to a moderate increase in part-time farms (Reimund, 1982). Such an increase suggests that these farms will never grow to provide a career for farm family members. Rather, they have a secondary economic interest in farming due to the fact that their primary activity is in other sectors. Furthermore, in this context the low returns of small farms may not always be indicative of low family income, given the fact that a large portion of this income comes from off-farm activities (Chantfort, 1982).

The increase in these two opposing poles of the American agricultural structure is accompanied by a slow reduction of the number of mid-size family farms. Several reasons have been given to account for such a trend. First, the inflation of the late seventies and early eighties may have artificially boosted a number of medium-sized farms into larger categories. Second, some farms may have expanded their operations to a larger size or reduced them. Finally, the economic crisis and related foreclosures have forced some out of the business altogether (Reimund, 1982). In general, then, given the large amount of capital required for a full-time commercial operation to enter the sector (estimated at more than $500,000), the condition of saturation of the domestic and international market, and the increasing desirability of rural over urban life, it appears that we should expect a further consolidation of the dualistic structure of American agriculture (Brewster et al., 1983; Chantfort, 1982; Lin, Coffman, and Pen, 1980).

West Germany

West Germany is characterized by a sharp contrast between its industrial and agricultural sectors. Industry is very developed, extremely productive, and quite profitable. Agriculture, on the other hand, is "inefficient in its use of land, capital and labor" (Mayhew, 1970:54). The poor performance of German agriculture has been attributed to its dualistic structure (Clout, 1975). The small farm sector makes up over 54 percent of all farms but controls only 16.8 percent of all land utilized. The large farm sector, on the other hand, is composed of only 2.9 percent of all farms but controls almost 17 percent of land utilized and accounts for over 30 percent of all agricultural production. The condition of the small farm sector, which is the result of pre-capitalist farming practice and laws of inheritance (Clout, 1975), is more precarious than in appears at first examination. In fact, 44 percent of all German farms have a size of 5 hectares or less, and farms under 20 hectares comprise 83 percent of all units.

The physical size of the majority of the farms is not the only problem. As Clout stated: "Many of the holdings are fragmented into widely scattered parcels of land and farm buildings are packed into villages and separated from the farm proper, making the efficient use of labor and capital even more difficult" (1975:180). The problems related to the presence of a dualistic agricultural structure are aggravated by the distribution of small farms. There is a concentration of small holdings in a few regions of Germany such as Rheinland-Pfalz, Baden-Wurtemburg, Western Bavaria, and Hussen. In these regions the relative importance of agriculture is greater than that in the rest of the country due to the lower level of economic alternatives available for the local population. First, the percentage of workers engaged in agriculture is well above the national average (Mayhew, 1971). Second, farm incomes are major sources of support for local families. Finally, farm incomes are substantially below industrial ones and generally poor in absolute terms (farms incomes are equal to 25.8 percent of industrial ones) (Coult, 1975).

In West Germany, then, there is a large farm sector that is very productive but extremely limited in size. There is also a small farm sector that is less efficient but contains a substantial portion of farms whose owners often have limited job alternatives outside of farming.[6] Attempts to correct this situation have been made. Land consolidation programs have been adopted since the middle fifties, but they were more effective in regions in which the presence of small farms was not very great (Clout, 1975). In addition, efforts have been made to encourage farmers to leave the farm. Industries have been attracted to rural areas and, as a consequence, there has been a proliferation of part-time farms. Simultaneously, an increasingly large amount of land has been left uncultivated or planted out to fruit trees or timber. These efforts, however, did not seen to alter dualism but only to limit it or to generate qualitative changes with the proliferation of part-time farms.

Spain

Spanish agriculture has undergone a strong process of development during the last twenty years that has largely overcome is outdated traditional traits. This change was generated by two major events. The first was an intense process of rural out-migration. This started tentatively during the fifties but reached dramatic levels during the sixties, profiting from the industrial development that took place in major European countries and in Spain's urban areas where the migration flow was directed. Within only a few years in the early sixties over five million farmers left agriculture, and ten years later the number of farmers was further reduced by two million in a country in which the total population does not exceed forty million (Perez Blanco, 1983). The second event was an increase in the standard of living related to industrialization, which stimulated the market and generated a larger demand for agricultural products (Garcia De Blas, 1983).

In spite of these changes, however, Spanish agriculture is still rather poor when compared with other Western countries. The major reason for such a situation is that a large portion of farms are too small to reach satisfactory levels of production, and among these a substantial number are marginal (Benito, 1982). Policies of consolidation aimed at enlarging the size and increasing the productivity of farms have been undertaken in Spain as well. However, their success was limited by the impossibility of further reducing farm population. In fact, since the middle seventies the channels of out-migration both domestically and, above all, internationally were interrupted due to the economic crisis. On the other hand, the timid process of rural and domestic industrialization was delayed quite substantially as a consequence of industrial restructuring, leaving an abundant underemployed population on Spanish farms (Perez Blanco, 1983).

The role of a reservoir of labor played by Spanish agriculture is stressed also by the fact that almost 16 percent of the population employed in agriculture are individuals under the age of 25. This figure is only 2 percent less than that of young workers employed in Spain as a whole (18.2 percent), and it is almost twice as large as that of young agricultural workers in member countries of the European Community (9.4 percent). The most recent data available show also that over 68 percent of all land is occupied by farms that generated an annual income of $5,000 (700,000 ptas.), and another 12.4 percent of land is occupied by farms that provide an annual income ranging between $5,001 and $7,500 (less than 1 million ptas.). From the geographical point of view, it must be noted that a substantial portion of these farms are concentrated in only two regions, Andalucia and Extremadura, complicating even more the attempts of structural reform conducted by the Spanish authorities.

The large farm sector, which is rather limited, is the more dynamic part of Spanish agriculture. In past decades these farms were objects of substantial investments aimed at both improving the structural conditions and mechanization, as well as technical inputs (Perez Diaz, 1983). The formation of this sector arose from the improvement of some small and/or medium-sized farms and from the breakdown of latifundium farms that obtained an advanced productive organization through restructuring. They cover about 10 percent of the land utilized and account for 35 percent of the total agricultural production of the country. Moreover, their rate of productivity is growing faster than those of similar farms in other Mediterranean European countries such as France and Italy. The large farm sector also seems to receive considerable attention from the government. In recent years (1971-1982) Spain's agrarian policies have gradually changed from those emphasizing the reform of small holdings as a whole to those stressing modernization and technological improvement of farms already at levels of adequate efficiency (Gomes, 1983). Structural dualism here is more accentuated than in the other countries examined above. The two farm sectors are very different entities with differing problems, and they are moving in different directions.

France

France's economic structure is characterized by regional concentration (Clout, 1975). In the Paris region are found not only the most important administrative and informational centers but also the largest industrial aggregates and the most advanced and productive farming units, which together generate the highest per capita income of all Europe (Clout, 1975; Klatzmann, 1978). The centralized character of France's economy also affects the agricultural structure, and only a few other regions contain farms that can reach production and income levels near those of the "Ile de France." In fact, there exists in France a significant mid-size sector that covers 53 percent of all farms and 51 percent of all land. The small farm sector (farms under 10 ha) comprises 35 percent of all farms and almost 7 percent of all land, while the large farm sector occupies 42 percent of all land and accounts for 12 percent of all farms.

The overall relevance of agriculture in the country is also diminished. In the last decade the percentage of labor force occupied in the sector decreased from over 13 percent to 9 percent, while the contribution of agriculture to the GNP fell from 6.3 percent in the early seventies to 5 percent today. Nevertheless, within French agriculture there exist sharp differences at the productive, occupational, and regional levels. First, most of the mid-size farms cannot provide an income sufficient to support the farm family. Consequently, in recent years a considerable portion of these farmers have taken an off-farm job (Klatzmann, 1978). Today, part-time farms constitute 65 percent of all farms. Only 25 percent of farms are run exclusively by family labor, 3 percent use a combination of hired labor and family labor, while the remaining 7 percent use only hired labor. Second, almost the entire totality of small farm families receive welfare payments which represent an increasing portion of their income, while large farms provide rising returns to their owners (Klatzmann, 1978).

Finally, the regionalization of such an economic differentiation is evident. In the regions where small farms are concentrated, especially Central and Western France, the average farm income is under 20,000 francs (1 US dollar = 7.0 francs; 1986 values) while in areas where there is a concentration of larger farms the annual average farm income is above 100,000 francs (Chombart De Lauwe, 1979). Moreover, in the regions where small farms are more concentrated, there is a lack of job alternatives outside farming that impedes, on the one hand, the possibility for these farmers to leave agriculture (this option is also limited by the industrial crisis in core areas) and, on the other, to remain in farming as part-time operators. As far as agrarian policy is concerned, the efforts of the French authorities have been addressed towards two major groups of farms, those that already have a high level of productivity and those that display a potential for modernization (Klatzmann, 1978). Small farmers have been encouraged to leave agriculture and have been the subjects of policies of social assistance. Small and large farms in France are, thus, two opposite faces of an agricultural sector that is farm from being homogeneous in spite of its appearance.

To be sure, dualism is not the absolute norm in the agricultural structure of advanced western countries. As mentioned above, Great Britain together with the Netherlands, Belgium and Denmark are notable and important exceptions. In Great Britain large farms constitute the largest segment in the agricultural spectrum, constituting over one-third of all farms and occupying 82 percent of all available land. The Netherlands, Belgium and Denmark have agricultural structures characterized by a large middle sector and relatively limited large and small farm sectors.

To recapitulate, there are conceptual and empirical problems related to the analysis of the concept of small farm at the international level. Nevertheless, it is evident, despite some important exceptions, that structural dualism is a major characteristic of western agriculture. Its importance lies in the fact that the two sectors are very different entities facing differing problems at various levels, yet they are unified under the same national economic system and the same international mode of production. Chapter 3 focuses attention on Italy. It provides a review of the major historical events that shaped Italian agriculture to its present dualistic form and identify the characteristics and the trends of the large and small farm sectors.

3
Patterns of Agricultural Development in Italy

THE DEVELOPMENTAL STRATEGY APPLIED IN ITALY SINCE WORLD WAR II

Most of the socio-economic literature concerning Italy begins analysis at the end of the Second World War. The selection of such an historical period is due to the simultaneous occurrence of three funda- mental changes. The first, which took place at the institutional level, involved the creation of the Republic with the proclamation of a new constitution endorsing the fundamental rights of a modern democratic State. The second change took place at the political-economic level and involved the end of the fascist regime and the entrance of Italy into the Western bloc under the influence of the United States. American interests began to play a dominant role in Italian affairs, especially with respect to Italy's position in the international arena. The last change occurred at the social level and had to do with the collapse of the former dominant social bloc that Gramsci (1974) described as the alliance between the Southern agrarian aristocracy and the Northern industrial bourgeoisie. It was replaced by a new alliance between the industrial bourgeoisie and the State bourgeoisie, a new class fraction in charge of administering the increased intervention of the State in the economy (Graziani, 1979). These changes did not alter, however, the sharp differences between the Northern and the Southern regions of the country. The North was developed with a strong industrial network, good agriculture, and infrastructure second to none in Europe. The South was underdeveloped, poorly industrialized, and with an obsolete agricultural sector dominated by latifundia, which were large estates owned by absentee landlords. This North-South dualism has been a characteristic of Italy since the unification of the country in 1861 and, in spite of numerous attempts made to eliminate this gap, it has grown steadily.

Favorable international and domestic conditions enabled the Italian socio-political establishment to select a developing strategy based on the expansion of some branches of mechanized industry such as automo- biles and home appliances. According to defenders of this strategy, the intermediate technologies of these industries and the relative labor intensity of their processes would not only have an effect on profits and the rate of national growth, but they would also relieve the entire country of the heavy post-war crisis (Mutti and Poli, 1975).

The socio-economic context within which this strategy was to be applied was complex, and the position of the major social classes can be summarized as follows. The working class, both in rural as well as in urban areas, was paying the consequences of the post-war economic crisis with unemployment and destitution. In urban areas unemployment was due to the destruction of a large part of the industrial network during the war. In rural areas, and especially in the South, peasants and farmworkers seemed to have no occupational alternatives outside of agriculture (Daneo, 1969; Mottura and Pugliese, 1980). After twenty years of a Fascist regime, characterized by an anti-peasant agrarian policy (Rossi-Doria, 1979), the economic conditions of the Italian rural population had not improved over those present at the end of World War I. Moreover, Mussolini's policy of increasing the size of the rural population, which was pursued through the use of economic incentives for large families and the prohibition of rural migration to urban areas, led to an unbearable rural overpopulation. Under these conditions the rural masses' only possibility of improving their conditions was to undertake actions aimed at breaking up the latifundia and obtaining a redistribution of land. The primary goal of the strong peasant movement formed in these years became that of agrarian reform, which was pursued intensively and with a heavy cost in human lives.

The Italian bourgeoisie, however, sought to accomplish three goals. The first of these was that of reorganizing the industrial apparatus in order to regain the margin of profit lost due to the war. To achieve this goal it was necessary for the industrial bourgeoisie to break the power of the backward Southern Italian aristocracy, namely, to end the latifundium. The presence of a latifundistic structure of agriculture represented a barrier to the development of more advanced agricultural units. It also inhibited a more rapid circulation of capital because of less investment and less profit. Moreover, it did not allow the enlargement of a domestic consumer market for industrial goods, since thousands of peasants limited consumption to products produced on their own farms (Graziani, 1979). As mentioned earlier the end of the latifundia was also the goal of the peasant movement, which provided still another reason not to defend it actively.

The second goal of the bourgeoisie was that of legitimizing its social hegemony in the country (Mutti and Poli, 1975). The political and social pressure exercised by the working class was increasing considerably under the direction of the Italian Communist Party and the Unified Trade Unions, which demanded an effective participation of this class in the direction of the government. Of course, the participation of the working class in the government (through the presence of members of the Communist and/or Socialist party in the Cabinet) was a serious obstacle to the realization of the reconstruction strategy, the stability of the political system, and Italian membership in the Western alliance (Podbielski, 1974).

The final goal was that of controlling the large labor supply available in the country (Mottura and Pugliese, 1975). The process of industrial reconstruction, which was taking place at a moderate pace, did not allow for the full deployment of such a labor force immediately. Unemployment was also the source of major social tensions, so it was important to somehow keep this cheap labor force crystallized as it was until the time was ripe for its integration into the new economic system.

The part assigned to the agricultural sector in the strategy of development was twofold. First, from the occupational point of view, agriculture, and in particular Southern agriculture, was the sector in which the above-mentioned large supply of labor was conserved (Mottura and Pugliese, 1980). Second, from the productive point of view, agriculture was considered to be a secondary sector, since the margin of profit that it returned for units of capital invested was considered less satisfactory than that derived from industry. However, the incompatibility of the latifundium with the new strategy of development led the industrial bourgeoisie to favor the creation of new and more advanced farms. In turn this meant strengthening the still weak agrarian bourgeoisie and limiting the burden of a disappearing yet strong class of latifundists.

In summary, the characteristics of the initial post-war period in Italy were:

1) A strategy of development that favored industry, especially light industry, for the production of durable goods with an intermediate technology in the hope that this might lead Italy to a strong and rapid economic development. Agriculture had a secondary productive role because it was the sector that contained the excess labor force.
2) The working class suffered the brunt of the crisis in both urban and rural areas. Furthermore, the deplorable socio-economic conditions of the period generated a wave of struggles for better living conditions that were embodied in rural areas in the struggle for an agrarian reform.
3) The bourgeoisie had three major strategic goals in its policy of reconstruction: a) to reorganize the process of accumulation through the reconstruction of the industrial apparatus and the elimination of the backward agrarian aristocracy from the dominant social bloc, b) to calm the social tensions in the country and to legitimize its hegemony, and c) to control the large supply of labor.

STATE INTERVENTION IN AGRICULTURE IN THE FIFTIES

The situation in agriculture at the beginning of the 1950s was characterized by high unemployment and the presence of a large number of backward farms and latifundia in the South. The intensity of the peasant struggle and the desire of the domestic bourgeoisie to break the power of latifundists and to obtain political support from the rural masses generated the conditions for agrarian reform. Some 742,500 hectares of land were expropriated under the law of Agrarian Reform (called "Stralcio" or "extract" as it was an extract of a more comprehensive law that never passed), but the land assigned to peasants amounted to only 681,581 hectares of which 455,828 ha were assigned in the South (the remainder of the land expropriated was either returned to the latifundists, as indicated below, or improved and sold by the State to the highest bidder). The latifundists had the right to receive payments in the form of State bonds in return for their expropriated land. they also had the option of requesting one-fourth of the total amount in cash if they agreed to improve the condition of the land left to them. Moreover, the latifundists expropriated could keep one-third of their land if, again, they agreed to improve the land themselves. Peasants were entitled to receive land if they were actively working in agriculture, the head of a family, and did not already have any parcel of land large enough to support a farm family. The peasant had to pay for the land within thirty years and could not sell or rent it to anyone for any reason. The land involved in the Agrarian Reform included 29 percent of the total available for agricultural purposes in the entire country.

According to the 1951 Italian census (ISTAT, 1952), 133,066 families received land (79,795 in the South) for a total of about 500,000 people of of eight million agricultural workers present at the time in the country. The average size of the farms created was six hectares (13.2 acres). It was lower in the South (5.7 ha) where, due to the hilly nature of the ground, the size of the farms needed to be larger than in the rest of the country in order to obtain sufficient economic results. The ratio between the number of peasants who received land from the reform and the amount of land given was that of two hectares per person (4.4 acres).

The quality of land was also not high. In many cases it was marginal with a low potential for good quality and abundant crops. Furthermore, there were no housing facilities for the peasants on the land, and the program for rural housing sponsored by the central government turned out to be an unsurpassed fiasco (Pezzino, 1972).

Although the agrarian reform did not improve the conditions of the peasants (Fabiani, 1979; Mottura and Pugliese, 1975), it did generate several important changes in Italy. As Mottura and Pugliese wrote:

> The first and most immediate (change) was the breakdown in the unity of the labor front created in rural areas during the post-war period, which meant the isolation not only of farm laborers but of the industrial labor class as well. The second... was the crystallization of the principal share of relative surplus population in the form of those agriculturally employed in enterprises

> dependent in large measure on the supply of public financing. The third was the possibility for moderate forces in the government to make a solid base of political consensus out of such strata of farmers by means of the spreading of capillary organizational structures modelled along corporate lines, tinged with technical efficiency and given the power to regulate the supply of public funds (1980:184).

In fact, the majority of the farms created by the agrarian reform were very much dependent upon financial and technical aid from State agencies. Through these agencies the dominant class initiated a process of legitimation that led to the creation of a large consensus for the Christian Democratic Party in the rural areas of Italy, especially in the South.

Two of these State agencies, *Federconsorzi* and *Coltivatori Diretti*, played an important role. The former was the agency in charge of supplying technical devices and assistance to farmers. It operated at the national level and had a monopoly over technical assistance to small farmers. The *Federconsorzi* was also entitled to sell particular agricultural devices (such as pieces of machinery, fertilizer, etc.) at substantial discounts. Its action had a tremendous impact on the organ- ization of the agricultural sector. It opened the rural market to industries that produced durable goods and also guaranteed a steady demand for these goods. It also became a mean of controlling and shaping agriculture in the interests of industrial corporations (Stefanelli, 1968). The other agency, the *Coltivatori Diretti* (C.N.C.D.), was originally created as a Catholic trade union for farmers and was very closely linked to the Christian Democratic Party. The C.N.C.D. was very active in those years in organizing a network of activities that covered almost all aspects of farm life from technical to cultural to medical aid. In exchange the agency requested political support (votes in local and national elections) for the Christian Democratic Party (Fabiani, 1979).

The consequences of the agrarian reform for the latifundists were quite different. Paradoxically, it generated several positive transformations for the members of that social class. The funds that they received for the expropriations exceeded, in many cases, the real value of the land itself. At the same time, the reform stimulated the latifundists to change their economic attitude and become modern agrarian entrepreneurs. As a matter of fact, out of the total amount of land improved by the reform half of it went back to the former latifundists. With the capital received and the improved land, they then began to resume their activity in agriculture in a new and more profitable fashion. Investments in agriculture, however, were not the only utilization of the capital made available to ex-latifundists. A large number of them preferred to invest capital in urban activities that seemed more profitable. "These funds.. were directed outside the primary sector toward that particular sector of industry which during the fifties was undergoing a phase of general, heavy expansion (including employment levels)*, namely the construction sector" (Mottura and Pugliese, 1980:185).

A conspicuous source of investments in agriculture during those years was represented by the State itself. Besides the capital indirectly employed in agriculture through the agrarian reform, several acts were passed that involved other forms of investment in the sector. They took the form of land improvement programs such as land reclamation, irrigation, and the construction of infrastructure, which very much improved the general condition of the farming sector especially from the productive point of view. In fact, productivity increased everywhere at a very substantial rate, but it showed stronger improvement in areas of the South where State intevention was greater. In the Northern portion of the country, which contained the most advanced farms, there was an increase of productivity of about 50 percent in the areas affected by the agrarian reform. In the South this improvement was much greater. The whole region of Calabria recorded improvements of 100 percent (Bandini, 1956), while Sicily and other Southern regions performed about 90 percent. Production also rose at a faster rate in the South than in the North, 9.8 percent versus 5 percent, respectively.

CHANGE DURING THE SIXTIES

The years between the end of the 1950s and the beginning of the 1960s constitute a crucial era in the recent economic and social history of Italy (Mottura, 1975). During this time the country was transformed from a society based on rural activities to an urban, industrial-based society. The prime characteristic of this period was the

*Parentheses in the original

urbanization of five million farmers who left agriculture and migrated to the cities. This rural-urban switch was also a South to North migration, which provided an additional dimension to the phenomenon. The reasons for such massive migration can be summarized as follows. The reorganzation of the industrial sector, which began at the end of the war, reached a point at which Italian industries succeeded in combining a modern, efficient organization of production with the low costs of labor available in the country. These factors made industrial goods very affordable in a domestic and international market in which economic expansion stiumulated demand (Graziani, 1979).

From the occupational point of view, this economic growth, which was measured at an unsurpassed rate of 11 percent for 1959 to 1963, generated the mobilization of a substantial portion of the underemployed labor force that has been crystallized in agriculture and in the South in previous years. These workers were attracted to industrial employment in the North with salaries far superior to those to which they were accustomed.[8]

At the same time important events were taking place on the international level. The most relevant of these was the formation of the European Community, which was created at the end of the 1950s with the treaty of Rome. The principles of the EEC favored the creation of a free circulation of goods among the European countries so as to build an economic unity that would be the first step toward a political unity of all Europe. From the point of view of the European industrial bourgeoisie, the creation of the Community was desirable for several reasons. First, it provided the possibility of better control of the market and the end of economic protectionism within Europe (Fabiani, 1979). Second, during a period of industrial growth it was important to control the agricultural market in order to avoid instabilities in food prices that could result in social tensions in urban areas (Clout, 1975). Finally, there was the risk that an increase in food prices could generate a demand for higher wages from industrial workers and, consequently, a possible inflation and reduction of profits (Fabiani, 1979).

The creation of the EEC had strong support from the United States as well, for the American government saw Western Europe as a large potential market for many of its agricultural products. Fabiani (1979) pointed out that European imports of American agricultural goods increased by 161 percent in the first ten years after the European Community started, while the increases of imports from other countries increased only 55 percent. Additionally, American overall exports to the EEC increased from 21 to 52 percent during the same period.

Italy's interests in the constitution of the EEC were very much oriented toward the possibility of finding new and more adequate markets for industrial products. In spite of the domestic economic expansion, the local demand was not sufficient to absorb an industrial production that was growing at a much more rapid rate (Graziani, 1979; Rodgers, 1970). Moreover, Italy was also in favor of a free circulation of labor. The unemployed rural workers of the South could have job opportunities in other countries and through migration could ease the high demographic pressure of farms.

Given these priorities, little attention was paid to the basic problems of agriculture, especially those related to the productive and economic development of backward farms in Southern regions. Other European countries such as Holland, Belguim and France placed a strong emphasis on the development of agriculture as a condition for their membership in EEC, which prepared the way for further development of the sector in those countries.

Once created, the agricultural strategy of the ECC has been characterized by two different policies, that of prices and that of structures. The first has been a constant trait of the EEC since its creation, while the second was initiated in the late 1960s and developed in the seventies. The policy of prices refers to a series of price support programs aimed at guaranteeing farmers' remuneration in the event of a negative trend in market prices. However, as the tendency toward a relative reduction in agricultural prices has been a constant characteristic of the last decades, the EEC has been forced to use this policy each year. At the beginning of every agricultural year a special agency of EEC, the FEOGA (European Fund of Organization and Orientation of Agriculture), fixes a minimum price for each product. If the market price is inferior (as it always is) to the fixed price, the FEOGA will pay the difference to the farmer. The EEC has always justified this action as providing a guaranteed income to each producer and especially small producers. However, it has been pointed out (Mottura and Pugliese, 1975; Zeller, 1970) that this policy generates different consequences among the various types of farms and that large farms enjoy most of the advantages.

This point can be clarified using the example of two farms, one small and the other large. The production of each farm is normally related to its size in that the small farm produces less than the large farm. Moreover, the small farm tends to have larger costs of production per unit of product than the larger farm. If the market price is inferior to the fixed price, both farms receive the difference between fixed and market price from the EEC. The differences between costs and returns will therefore be higher for the large farm because of the lower costs of production, as well as the total amount of support received due to the larger quantity produced. Price support programs, however, are not applied indiscriminately to all the agricultural products of the EEC. In fact, the larger portion of the programs concerns dairy products and grains, which are typical of Northern European countries. Southern European products such as fruits and vegetables receive considerably less attention. Moreover, the policy of price created several distortions within the EEC. First, the artificial determination of agricultural prices forces European consumers into a situation in which they have to pay prices well above those available on the world market. Second, the costs of such a practice are born differently by consumers in various countries. In Germany, for instance, the part of the disposable income devoted to food consumption is 17.1 percent, while in Italy it is 34.5 percent and in Greece, 43.8 percent. This indicates that by creating inflation the policy of price reduces the buying power of Greek and Italian consumers more than that of their German counterparts (Clout, 1975; Fabiani, 1979). Finally, price support programs create surpluses. In order to eliminate surpluses, European countries are forced to acquire agricultural products from the EEC at the EEC's price, having been denied the possibility of buying cheaper products from non-EEC countries.

The second strategy that has been implemented in the EEC is the so called "policy of structure." This policy, added to the policy of prices, was embodied in the Mansholt Plan, which was written in the latter part of the decade. The Mansholt Plan, or Memorandum as it was called, dealt with two major problems created by the policy of prices. The first of these was the excessive cost of the policy of prices. The second problem involved the delay in the process of modernization of European farms. This was largely blamed on the policy of prices, which allowed the persistence of some small and backward farms. The memorandum's goals were, then, to eliminate the inefficiency generated by the previous policy and bring the income of farmers to the same level as that of industrial workers (Mansholt, 1968).

The attainment of these goals required a total reorganization of the structure of European agriculture, which involved the elimination of those farms whose operations failed to provide a sufficient income for the farm family. It also demanded a reduction of the farm population. To achieve these goals the memorandum suggested the following measures: a) giving financial support only to those farms that were already very efficient, b) establishing a set of operating procedures for the development of these farms, and c) developing a disengagement policy for those farmers who could not improve the economic efficiency of their farms and were near or beyond retirement age. The logic as well as the consequences of this policy were not very different from those of the policy of prices. In fact, it legitimized the unequal logic of the two agricultures, favoring large and more developed farms at the expense of less competitive and marginal ones (Saccomandi, 1978). The effects of the policy of structure, however, have been very limited in comparison with those of the policy of price, which has been the most important aspect of EEC agricultural policy since the 1970s.

The new international situation greatly influenced the domestic agrarian policy in Italy, which was characterized by the two "Green Plans" _(Piani Verdi)_ of 1961 and 1965, respectively. The results of the application of the two five-year plans deepened the already clear dualistic structure of Italian agriculture, as they provided generous aid to the large farm sector but added little to the small farm sector. The first of the two plans was the least selective. Its main concern was to regulate the shift of the labor force from the farm sector to urban areas (Mottura and Pugliese, 1975). This goal was attained through a series of benefits to small farmers which, allowed them to remain in operation a little longer yet did not change the structural characteristics of the farms nor improve their overall conditions. From the point of view of the owners of large farms, the first plan had two primary effects. First, the benefits for the formation of small farms (which were part of the plan) created an increase in land prices, which favored those who owned large estates. Second, the aid for increasing productivity accelerated the process of restructuring within the overall sector, which benefitted large farms the most.

The second Green Plan was more selective. It was designed to support those farms that were already competitive economically and technically. An example of such an expanded selectivity may be seen in the financial policy of the plan. In the first Green Plan financial support was granted with almost no collateral, and farmers who received aid did not have to repay any of the funds granted. In the second plan financing was based on loans that turned out to be more accessible to large farms due to their ability to provide collateral and eventually repay the loans. Of all financial aid under the first plan, 55 percent went to small farmers, while only 22.7 percent of that provided by the second plan went to farms in this category. Conversely, large farms received 79.3 percent of the funds available under the second plan versus 45 percent under the first plan.

STRUCTURAL AND OCCUPATIONAL CHANGES IN AGRICULTURE DURING THE SIXTIES

Italian agriculture at the end of the 1960s had changed a great deal since the beginning of the decade. Indicators of the change include: a) a sharp decrease in the number of farms (from 4.2 million to 3.6 million) b) a 3.5 million hectares reduction in cultivated land; c) a decrease in the agricultural portion of the GNP from 23 percent in 1961 to 9 percent in 1971 and d) a decrease in the number of workers employed in the agricultural sector (from 5 million to 3 million). At the structural level, there was a strengthening of the dualistic composition of Italian agriculture. The decrease in the number of farms included only medium and small farms, though the latter still represented the largest portion of all farms. Large farms gained both in numbers and in size, with farms of more than 100 hectares gaining 670,000 hectares as a whole. This gain came only a few years after the Agrarian Reform in which a similar amount of land was redistributed to small farms.

At the productive level there was a substantial increase in both productivity of labor (6.4 percent) and in productivity per hectare (2.9 percent) which reached the level of the most advanced European agriculture. However, there was also a decrease in the rate of growth of production that was 3.2 percent at the beginning of the decade and fell to just under 1 percent at the end of it. Consequently, Italian agriculture was increasingly incapable of satisfying the food needs of the country and had to rely on substantial imports from abroad. The deficit of the agricultural balance of payments rose from 132 billion lira

($130 million) in 1961 to 400 billion lira ($3,800 million) in 1974. Furthermore, the increase in productivity of labor and productivity per hectare applied almost exclusively to the large farm sector, whose expansion was not able to make up for the lack of production created by the reduction of the small farm sector (Fabiani, 1979).

In the 1960s North-South dualism and the dualism between the hilly, mountainous areas and the plains were accentuated. As mentioned above, the contribution of agriculture to the GNP was reduced to 9 percent. This figure varies, however, from region to region. In the North it accounted for 5 percent of the GLP, while in the South is was 18 percent. This difference meant that in the North agriculture was not a very important economic activity, since it comprised only one-twentieth of the wealth created there. However, in the South it was very important, since it provided one-fifth of all income. Moreover, in that region the sector that generated the largest part of the GLP was Public Administration (local and governmental agencies), which did not represent a real economic sector but redistributed wealth accumulated elsewhere and was collected by taxes (Pugliese, 1977). Accordingly, in the South the relative importance of farming was higher than it appears.

As far as the dualism between hilly areas and plains is concerned, important changes took place during the decade. In the plains, including the South, large farms grew in number. Also, those areas received the largest portion of private and public investments made in the sector. In the mountains and hilly regions, on the other hand, there was a decrease in investments and an increase in the concentration of small marginal farms. From the occupational point of view, the basic change during this decade was the decrease in population employed in agriculture. In absolute value the largest decrease took place in the South, but in relative terms it was higher in the North. The decrease of employment in agriculture was positively correlated with the availability of job opportunities in other economic sectors (Paci, 1973). In the North the availability of job opportunities outside agriculture reduced the farm population, permitting the restructuring of a significant portion of farms and the further productive development of the sector as a whole (Pugliese and Rossi, 1975). In the South, due to the lack of job alternatives, agriculture still maintained an important role as an occupational resource. However, the employment provided was increasingly precarious. There was an increase of farm workers and a decrease of independent farmers. Moreover, out of a total of 1.2 million farm workers, only two-hundred thousand had steady work year round. The others worked for a minimum of 50 or a maximum of 200 days per year (people who work for less than 50 days per year are not considered agricultural workers by the Italian Census Bureau (ISTAT) and, consequently, they are not included in available statistics). Even in this case the majority of periodically employed farmworkers were concentrated in the South.

THE TRENDS IN THE SEVENTIES

The 1970s witnessed the development of a profound economic crisis, affecting the entire industrialized world. As a consequence of this international crisis some changes occurred in the EEC's agrarian policy. EEC payments prior to 1973 were made on the basis of the value of the US dollar convertible into gold. A "unit of exchange" was selected as European currency and tied to the value of the dollar so that exchange among the various European countries occurred at the same rate. This procedure facilitated the EEC's direct intervention in the economy (e.g., price supports, financing, etc.) and also made it possible to control easily agricultural prices among all of the EEC countries.

In 1973 the end of the convertibility of the dollar into gold threw the entire system into crisis. As a result, the EEC was divided into two currency zones: the so-called Mark area, which was connected to the Deutschmark, and the Dollar area. The former included Germany, the Netherlands, Belgium and Denmark, while the latter was made up of Great Britain, France, Ireland and Italy. The currencies of the countries in the Mark area were revalued, while those of the countries in the other area experienced a steady depreciation. In terms of agricultural prices, this new situation meant that in the Mark area there was a reduction of agricultural prices, while in the Dollar area there was an increase. Thus, the system of exchange within the EEC and the mechanism of price control in each country were compromised.

The decision was made by the EEC not to use the new rate of exchange within the European agricultural system, and EEC authorities agreed on a partial and gradual adjustment of exchange rates. In this way new rates of exchange (the ones determined by the market) would be fully applied within the European agricultural community within a specific amount of time, though not immediately. This strategy constituted an attempt to gradually move from the old monetary system to the new one without creating new contradictions.

This policy, however, left the problem unsolved because there were still differences in agricultural prices throughout the Community (Fabiani, 1979). A compensatory mechanism was then designed that taxed the exports of those countries with an inflationary economy, while the imports of countries suffering from the devaluation would be subsidized. The actual result of this policy was that relatively weak economies such as Italy saw their agricultural products penalized when exported, while relatively strong economies such as Germany had their agricultural trade financed by funds from all countries within the EEC. This system might have been a solution for a short period, but only in 1984 have concrete steps been taken for its replacement and not without great controversies and potential for a further delay.

This policy had several consequences for Italy. First of all, although it created a general disadvantaged for all agriculture, there was a certain sector of Italian agriculture that gained from it. The rate of exchange applied to Italian agricultural products within the EEC was lower than the market rate. Moreover, agricultural prices increased proportionally more than other prices, while the rate of exchange between agriculture and other sectors decreased somewhat. Consequently, there was an overall increase in the rate of profit in agriculture. However, given the distinction between large and small farms that characterizes Italy, only the large farms could take advantage of this situation due to their more extensive use of technical inputs and their tendency to export (Fabiani, 1979).

At the structural level, then, there was a repetition of the trend developed in the previous decade. During the period from 1970 to 1980, there was a decrease in small farms consisting of less than 20 hectares. These farms still made up 85.5 percent of all farms but only 37 percent of all land. At the same time those farms with more than 100 hectares decreased in number, but their average size increased so that the overall tendency toward the growth of large scale agriculture was confirmed (Fabiani, 1979).

At the occupational level, from 1970 to 1979 there was a decrease of more than 700,000 in the population employed in agriculture. This reduction was distributed very unevenly throughout the country, with a decrease of 500,000 in the North and the remainder in the South. This trend has been interpreted not as supportive of the hypothesis that agriculture can provide more job opportunities in the less developed South but, rather, that the lack of job opportunities in other sectors kept people from leaving agriculture (Fabiani, 1979; Furnari, 1980).

The rate of productivity increase also fell from a 3.0 percent in the 1962 to 1969 period to a low 1 percent in the period 1970-1979. Moreover, in the first half of the 1970s net investments in agriculture were quite low, even below those of the 1950s. Only after 1975 was there a reversal of this trend.

At the same time there was an increase in the deficit of the agricultureal balance of payments. The decreased ability of small farms to provide income for the farm family led to an increase in welfare benefits paid by the State. According to recent analyses, in the last few years Italian farmers and particularly Southern farmers have become heavily dependent on welfare payments, which make up over 70 percent of the total income of the farm family in the more extreme cases (Pugliese, 1983, 1977). Consequently, an increasing number of small farmers can no longer be considered agricultural workers as they have depended largely on the State for subsistence. In summary, in the period from 1970 to 1980, the following phenomena occurred:

1) A deepening of the dualistic character of the agricultural structure. Large farms increased in average size, while small farms decreased but still account for a relatively large portion of the farm sector.
2) There was a territorial differentiation within agriculture. Large farms tended to be concentrated in flat and more fertile areas, while small farms were located in mountainous and hilly areas.
3) The new EEC agrarian policy did not improve the condition of the sector but contributed to the widening of the gap between the two major groups of farms.
4) The agricultural productive capacity of the country was reduced and, in spite of the declining demand for food-stuffs, the need for imports increased.
5) An increasing number of small farms, especially in the South, could not provide an adequate income for the farm family. These farmers were thus forced to rely on welfare benefits from the State for their sustenance.

Chapter 4 introduces the central issue of the present work, that of the persistence of small farms in South Italy, by providing some general information about the methodological aspects of the research and, secondly, by describing some of the major socio-economic characteristics of South Italy and the area surveyed.

4

Underdevelopment and Marginal Farming in Southern Italy

A BRIEF METHODOLOGICAL NOTE

The research pivots around four major concepts: Persistence, Advanced Western Societies, Marginal Areas and Small Farms. It is relevant, therefore, to state the definition of the main concepts adopted in the research and the procedures involved in the selection of the sample.

Persistance. This concept indicates the continuous existence of small farms within the capitalist mode of production. Operationally, it means the actual existence of small farms in Southern Italy and Sicily.

Advanced Western Societies. Advanced Western societies are countries that exhibit the culture and social structure typical of western civilization and more specifically the culture and structure typical of Western Europe and North America. The use of the term refers to the fact that socio-economic standards in these countries are among the highest in the world. Italy is the country to which we refer in the course of research.

Marginal Areas. This concept refers to regions that, in contrast with the rest of the country in question, are characterized by low per capita incomes, high levels of unemployment and underemployment, low levels of quality of life, and low contributions to the formation of the GNP.

Small Farms. Small farms in Italy are those with an annual product sold in 1981 of less than 20 million lira (at the 1981 value 1 US dollar = 1500 lira) and which occupy a land area of more than 1 hectare.

The universe of the research is represented by small farms located in the counties of Eastern Sicily and South East Calabria (Map A). Given the high number of small farms present in the two regions, a sample of small farms from each region was selected. The selection of the sample was based on the following procedure. Eastern Sicily is defined as the region occupied by the districts (provinces) of Messina, Catania, and Siracusa. In the three districts there are 325 counties (Counties in Italy are substantially smaller than those in the United States). Sixty-five of these were randomly selected (20 percent of the total), and 100 farms were selected out of the 65 counties according to the following procedures:

1) Using census of agriculture data, the percentage of small farms present in each selected county was computed out of the total of small farms present in the 65 counties selected.
2) To each county a portion of the sample was assigned which was equal to the percentage of small farms present in the county out of the total number of small farms present in all the counties according to the census. (For example, if the percentage of small farms in a county is 15% of the total of small farms present in the 65 counties selected, 15 farms will be selected from the county).
3) The actual selection of the small farms was carried out using the membership list of the Coltivatori Diretti and Alleanza Coltivatori, which are both trade unions for farmers.
4) After the farms were selected, each farm family living on the farm was contacted by telephone and preliminary questions were asked in order to establish whether the farm qualified according to the definition of small farm.
5) The farms that did not fit the specific criteria were replaced with other randomly selected farms from the same list. Identical procedures were used to select farms from the Southeast Calabria region, which occupies the districts of Reggio Calabria and Catanzaro.

At the end of the sampling procedure, 199 valid cases were obtained, and they constituted the body of data used in the research. A random selection procedure was used in order to have a clearer picture of the farm sector in the regions chosen, which could contain all the various types of small farms. The use of the membership list made available by the Alleanza Coltivatori and Coltivatori Diretti may create some bias with respect to farmers who are not members of either of the two trade unions. However, the two organizations include over 80 percent of all small farmers in both regions thus constituting a very large portion of the total universe. It was not possible to include the remaining 20 percent of small farms in the sample.

The unit of analysis for the present research is the farm family. This choice is based on the multiple contributions that family members provide for the survival of small farms. As it has been stated in recent studies (Coughenour and Swanson 1983; Pugliese, 1983), the farm family is preferable to other units of analysis such as the farmer head of the household or the farm as economic enterprise.

The data were collected using personal in-depth interviews with members of the farm family. After the first preliminary contact, the respondents were contacted again by telephone or letter and a date for the interview was arranged. The interviews were conducted on the farm or at the town house of the farm family by a researcher. After being authorized by the respondent, the researcher used a tape recorder in order to record the entire conversation. A written record of the interview was also made. The interviews, while generally conducted on an informal basis in the attempt to obtain the most reliable information possible, followed a pre-arranged schedule.

The data collected encompasses observations arranged according to statistical procedures and field notes arranged according to observational studies. These data were analyzed both qualitatively and quantitatively through the use of the dialectical method. In this method the statistical arrangement of data provides only supporting, rather than conclusive, evidence. This posture reflects the fundamental understanding in dialectic that statistically arranged data represent an epistemological and heuristic device that illustrates social trends, but is not sufficient by itself to establish sociological laws. In this respect history becomes the setting within which social action takes place and which, in turn, shapes social action.

As far as qualitative data are concerned, their use in the dialectic method differs from that of orthodox studies in three in stances. First, qualitative data are combined with quantitative, thus overcoming the rejection of the use of statistics in sociology which is typical of most qualitative designs. Second, the qualitative portion of the dialectical method is not based on the notion of "value freedom" (Weber, 1949). That is, the observer is considered as bearing political meanings in his/her action of investigation, so the neutrality of a vaule free posture is rejected. Finally, the qualitative analysis is not confined to the micro sphere but can be translated into the macro level of analysis. This process is possible due to the overall historical framework within which the social action is considered to take place.

Given this methodological background, it is the intention of this work to provide some social and historical information on the major characteristics of the region in question in order to identify its marginality with respect to the rest of the country and to stress the most important socio-economic trends relevant for the question of persistence.

SOUTHERN UNDERDEVELOPMENT AND THE SOUTHERN QUESTION

The origin of Southern underdevelopment must be traced back to the middle of the last century, when Italy was unified into one country. Before that time the peninsula was divided into several sovereign and independent States. Some were located in the North-Central part of the country, while the entire South was under the jurisdiction of the Kingdom of Naples, a State controlled by a cadet branch of the Bourbon family. At the time, there were already some differences in the level of development between the North and South of Italy. However, they were not as great as those recorded only a few decades later. The economic policy of the Bourbon administration was protectionistic. It allowed the slow but gradual development of an industrial network, which was mostly located on the coastal areas of the South (Gerschenkron, 1962). Nevertheless, agriculture was the major economic sector in the South.

In 1861 Italy was unified under the reign of Victor Emmanuel II, the former king of Piedmont. The unification meant, among other things, the end of the protectionistic economic policy applied to the South. In fact, the first Italian government (1861-1877) pursued a strong laissez-faire, free trade policy. The adoption of this policy was dictated by the desire of the Italian bourgeoisie to create a sufficiently large domestic market and to export agricultural products abroad (Mutti and Poli, 1975). This goal meant the elimination of any obstacle to the free circulation of commodities and, consequently, of any protectionist barrier within the country and with other nations (Sereni, 1968). It followed that the abolition of protectionism both domestically and internationally exposed Southern industries to the double competition of Northern and foreign enterprises. In only a few years the weak yet developing industrial network of South Italy rapidly disappeared.

As Gramsci (1973) pointed out, the economic policy of the first Italian government was not oriented toward the development of industry, but toward the development of agriculture. This was particularly true for the South. This economic strategy was due to the predominance of the rural classes over the industrial classes and the power that the landlords retained in the South. The weakness of the Italian industrial bourgeoisie favored the employment of foreign controlled capital, which was particularly large in sectors such as mining, railroads, and finance. Moreover, the importations of raw materials for industrial production created a large deficit in the balance of trade, which further weakened the international position of the country.

Italy, then, entered the international arena in a position of dependence with respect to the other European powers. Nevertheless, it experienced a constant process of growth and of financial rebuilding in the years following the unification. This process, however, was realized through the transfer of resources from the South to the North so that while the South was stagnating, if not regressing, the Northern portion of the country was rapidly growing. As Gramsci noted:

> The poverty of the South was historically inexplicable to the Northern popular masses. They did not understand that the unification of the country occurred on an unequal basis and as an hegemony of the North over the South. The North grew at the expense of the South, and its industrial economic development was in direct relation with the impoverishment of the Southern economy and agriculture (1974:92).

The destruction of the Southern industrial network was one of the reasons for the increasing gap between the North and South. However, other important events should be taken into account. First, the new tax system discriminated against the South. After the unification the Piedmontese heavy tax system was extended without changes to the entire country and, consequently, to the South were previously there was a rather light system of taxes. More importantly, though, the system was based on flat-rate taxes and taxes on consumption, so the poorer strata supported the heaviest burden. The majority of these people were concentrated in the South (Rossi-Doria, 1979). Additionally, State expenditures in the South (in terms of investments for infrastructures and other public activities) were substantially below the tax funds generated locally (Mutti and Poli, 1975).

Second, and in relation to the above, the unification of the economic and fiscal system of the country meant the unification of the national debt. Piedmont had a heavy national debt created by the wars that this small Northern Italian State had fought in order to secure the unification. The South, on the other hand, had a very limited national debt so that after the unification it had to face and pay a debt contracted exclusively by the North. Finally, there was a direct transfer of resources from the South to the North, which occurred in two manners. On one hand, former church and State properties located in the South were sold, and the funds obtained were reinvested in infrastructures almost exclusively in the North (Villari, 1966). On the other, the bank system collected funds in the South but systematically employed them in the North. Gramsci described this phenomenon as follows: "Through banks the savings accumulated with hard work in the South are invested in the North where factories exist and profits are high" (1974:9).

In 1878 there was a change in Italy's economic policy that brought the gap between North and South to a new historical high (Villari, 1966). From that time onward the Italian government of agriculture and increasingly supported the industrialization of the country.

Ten years later in 1887 Italy abandoned its free trade policy as well and entered a long period of hard protectionism. Such a change in the economic orientation of the administration can once again be traced back to the modified equilibria among the various fractions of the ruling social formation. Since the 1860s the power of the urban bourgeoisie (linked to heavy industry, steel and iron, and some branches of manufacturing) had increased considerably, while other fractions such as the commercial and financial bourgeoisie lost their position of predominance (Castronovo, 1975). Simultaneously, the rural bourgeoisie was never able to acquire any real power, leaving control of the countryside to the latifundist class.

The protectionist choice of 1887 and the "war of tariffs" against the major European nations, and more specifically against France, sealed the alliance between the Northern industrial bourgeoisie and the Southern latifundist class (De Rosa, 1973; Gramsci, 1974). Protectionism in the eighties primarily favored grain production, which was typical of latifundia, and the textile and iron and steel industries within the industrial sector. The protectionism of grain products created the conditions for the latifundists to survive the international agricultural crisis of the 1880s and, more importantly, to consolidate their power in the region. At the same time, it favored the industrialization of the North and allowed the industrial bourgeoisie to become the hegemonic fraction within the ruling class (Gramsci, 1973).

The consequences of this new situation for the South were devastating. The protectionist action against other European countries came at a moment in which Southern agriculture as a whole was becoming more and more export oriented. Products such as wine, fruits, and vegetables had found a growing demand on international markets that was drastically interupted, thus creating the conditions for a collapse of the network of small and medium-sized farms. Furthermore, the protectionism of grains sharply increased the price of bread, the traditional food for the Southern working class, which signified a worsening of the already low standard of living of this class (Mutti and Poli, 1975).

At the same time a process of unequal exchange was initiated between the South and the North (Capecelatro and Carlo, 1972). This process followed the patterns that have been described in the relationship between third world and core regions (Emmanuel, 1972; Frank, 1979), an exchange of more value (in this case from the South) with less value (from the North). In more specific terms it was characterized by the exchange of agricultural products from the South with industrial products from the North. As far as industry was concerned, it was characterized by the exchange of commodities produced through the use of less developed technologies with commodities produced with advanced technologies (Salvati, 1973).

The period from 1896 to 1908 witnessed the consolidation of the Italian industrial apparatus, which was accompanied by unprecedented economic growth. At the same time, though, the socio-economic separation between North and South was such that the two regions became almost opposite realities. There was development and capitalist organization of production in the North and underdevelopment and precapitalist system of production in the South. The condition of this dualism became so severe that the response of the Southern working class was a massive transoceanic migration to North and South America (Fua, 1969).

The First World War and the period between it and World War II were characterized by similar trends. Military expenses fostered first by the needs of the war and later by the imperialistic policy of the Fascist regime provided a further impetus to the development of the Northern industrial apparatus. However, during the 1930s the alliance between the latifundists and the industrial bourgeoisie entered its final stage (Castronovo, 1975). The social, economic and political costs of the maintenance of a latifundist class appeared too high for an industrial bourgeoisie pressured by strong international competition. The process of elimination of latifundists from the dominant social group would be completed only after the end of World War II.

After World War II the North-South relationship gradually began to change, and the development of the South became a constant concern for the ruling class. Efforts were made for its development both from an agricultural point of view, which culminated in the agrarian reform, and from an industrial point of view. The first strategy to industrialize the South selected by the Italian establishment was that of "infrastructuring" (Graziani, 1979). This strategy called for the creation of infrastructures in an underdeveloped region so that the conditions become attractive for investments that should spontaneously follow and trigger development. Such a strategy was derived from the modernization school, which was at its peak at the time and was represented by the work of scholars such as Rosenstein-Rodan, Nurske, Lewis, and Scitovscky. It was applied for over a decade from the early 1950s to the middle sixties but did not reduce the North-South gap (Graziani, 1979). However, some positive results were obtained (Graziani, 1979a). First, the infrastructure created attracted limited industrial investments in some selected areas of the South. These areas were those with an already existing industrial network rather than those in need of immediate economic growth (Boccella, 1982). Second, the policy of infrastructuring created as an indirect consequence a rapid process of urban growth, which generated the conditions for the existence of a network of small local enterprises and services (Del Monte and Giannola, 1978). Finally, the creation of infrastructure in the South represented an inversion in the traditional core-periphery relationship. In fact, in this case the flux of resources was inverted and flowed from the core to the periphery.

In the sixties this policy was abandoned due to its inadequacy in solving the problem of Southern underdevelopment. The new strategy adopted called for direct industrial investments in the South. These investments were carried on either directly by the State with public funds or were joint ventures between the State and large private industrial holdings. The investments had to be directed toward the creation of large plants within capital intensive industries. Steel and iron, automobile, and chemical and petroleum plants were located in selected areas of the South. The selection of only a few areas (instead of a more homogeneous distribution of these investments) was dictated by the strategy of the "development by poles," that consists of the identification of a specific area which is considered suitable for industrial development. Then industrial investments are carried on in the area which eventually would generate industrial development. Through the effects of economies of scale the development originally established in that area would spread to the rest of the region generating homogeneous growth.

In the specific case of South Italy, a few "poles" were selected, and in the original strategy their combined effect would have covered the entire region. Unfortunately, this strategy did not eliminate the North-South gap, nor did it improve the overall relative conditions of the South. It did, however, generate several changes in the structure of the South and in the North-South relationship. Among the changes that it fostered was the creation of a modern working class. Additionally, this strategy generated a clear differentiation between areas where industrialization took place and those where it did not. In the former the typical elements of developed regions existed, while in the latter the main traits were high levels of unemployment and underemployment, low per capita income, a high number of welfare recipients, and low activity rate. This situation suggested that social scientists now study "the Southern questions" in order to stress the differences and complexities existing within the region.

The presence of a modern working class in the South constituted an important difference between Southern underdevelopment and traditional underdevelopment, for the political power of this class is a primary importance in the region and in the country. Consequently, relevant goals have been achieved by workers in social and economic arenas. Since the middle 1960s wage levels have been homogeneous at the national level so that, in spite of the abundance of available labor in the South, wages remained high. Additionally, the Southern working class obtained large gains as far as welfare is concerned. Various types of benefits (unemployment, health care, education, etc.), were instituted for industrial as well as agricultural and service workers.

By the middle seventies the strategy of the poles of development was concluded. At the same time new processes of industrial restructuring were initiated. One of the most important of those was a spontaneous (spontaneous in the sense that it was not triggered by any developmental strategy from the State) process of decentralization. In this case tasks once performed by a single enterprise, and usually in a single plant, were decentralized to a number of small independent enterprises, which subcontracted with the original firm (Calza-Bini, 1976). This process of decentralization was particularly evident in Southern urban areas and was coupled with the other important phenomenon of the seventies, the growth of the "informal economy" (Bagnasco, 1981; Phall and Gershuny, 1982; Gershuny, 1979), which refers to economic activities that do not follow the orthodox forms of industrial production. They can range from cottage or household production to overtly illegal production involving the use of child and/or non-authorized migrant labor. The informal economy has grown so rapidly in the past ten years in South Italy that it has been estimated to contain about 25 percent of the local GNP today (Bagnasco, 1981).

The changes that have occurred in the last two decades have created a situation in which the character of Southern underdevelopment is today qualitatively different from that of the recent past. If fact, if the socio-economic situation in the South of Italy is compared with the traditional traits of underdevelopment, it must be concluded that the two differ markedly (Pugliese, 1983). First, underdevelopment calls for a transfer of resources from the periphery to the core, while in the present case this process has been reversed. Second, the cost of labor in underdeveloped areas is considerably lower than in core areas, as is the educational and technical levels of such labor. In the South of Italy the cost of labor is equal to that in the rest of the country, as well as educational and technical levels of labor (Graziani, 1979). Finally, underdeveloped regions lack influential representatives in the political arena. The political power of the South in the national context is, on the contrary, very high. Political representatives from the South hold very important positions both within the Cabinet coalition and within the two branches of parliament.

The South of Italy, then, is neither developed nor does it resemble "traditional" underdevelopment. It is dependent upon transfers of wealth from other parts of the country channeled through the State for its industrial investments, the survival of its industrial network, and for the maintenance of its large portion of welfare recipients. The South is poor as a whole, but within it some areas are well developed and resemble in their economic and social conditions the most advanced areas of Europe. The region surveyed falls within this complex framework. It is a marginal region that lacks industrialization and has a generally poor agriculture, but it contains a large tertiary network.

In the next unit of discussion a summary is provided of the most important structural characteristics of this region and of the farm sector as recorded through our survey and other secondary sources.

GENERAL CHARACTERISTICS OF THE REGION SURVEYED AND OF THE SMALL FARM SECTOR

The small farms surveyed are contained within an area that belongs to two different administrative regions: Calabria and Sicily. An administrative region within the Italian political system can be roughly equated to an American state, though regions have less political autonomy. The power of each is limited by the authority of the central government, which delegates some legislative and executive powers to the regional administration.

Calabria is among the poorest regions in the country. Its per capita income ranks nineteenth among the twenty regions of Italy, and its level of unemployment is the third highest in the nation. The major source of income for the local population is provided by tertiary activities, which account for over 50 percent of the local GNP. Among these tertiary activities the largest portion, over 60 percent, is represented by public and local government agencies. Agriculture represents the second major source of income, but only a few farms produce enough profit to guarantee a sufficient income for the farm family. These farms are usually large in size, though they employ only a limited portion of the abundant supply of farmworkers (56 percent of the available agricultural labor). Farmworkers are also employed by the forestry department for works of reforestation. Such jobs are more a relief for unemployment than related to an objective need for reforestation (Pugliese, 1983).

The industrial network in Calabria is also very limited, for the policy of the "poles of development" was particularly ineffective here (Graziani, 1979). Two major attempts to locate large plants in the region failed. The first involved the creation of a steel and iron plan in Gioia Tauro on the Southwestern coast of the region. Preliminary investments were made, and the construction of the plant was initiated but never finished. The international crisis of the steel sector and the financial inability of the Italian State to provide further capital and/or to reconvert the plant aborted the operation. The other "pole" was located on the Southeastern coast. It was a chemical plant specializing in the production of bioproteins, feed derived from mineral sources (generally oil). The plant was completed but never used due to the fact that bioproteins have been found to be carcinogenic. Under the pressure of lobby groups and concerned citizens the plant was closed indefinitely. No other large industrial plants are present in the entire region. Small manufacturing plants exist, especially in the textile sector, but they account for only a small portion of the total employment and local GNP. Industry employs only 28.8 percent of the working population, agriculture 28.7 percent and the tertiary sector 41.5 percent.

Sicily is almost twice as large as Calabria* and contains nine major urban areas versus the three of Calabria. However, from an economic point of view, the position of Sicily is not very different from that of Calabria. The region surveyed in Sicily includes the entire

*Sicily occupies 9,831 square miles, Calabria 5,822. The region surveyed in Sicily is about 1/3 of the total Sicilian area. The region surveyed in Calabria is roughly 1/4 of the total area of that administrative region.

eastern part of the island with three major urban areas. Within this area there are two industrial poles. One is in the Northwestern part of the island and the other is in the Southeastern part. The first pole, Milazzo, is characterized by several large petrochemical plants, while the second, Augusta-Priolo, contains petrochemical and manufacturing plants. These two areas have been selected for industrial development because of their favorable natural characteristics, flat land and an abundance of water. Both areas have a rather advanced agriculture with farms that produce adequate incomes for their cultivators (Nocifora, 1980). Part-time farming is also quite developed so that the industrial-agricultural apparatus provides job opportunities for the farm family (Chessari, 1980). However, these two areas represent a very small portion of all the territory surveyed and a smaller portion of the whole island.

The region surveyed is characterized by an urban network formed by three major cities: Catania, Messina, and Siracusa. Catania is the largest city of all with 400,000 inhabitants followed by Messina (251,000) and Siracusa (180,000). These cities are commercial and administrative centers which provide jobs for most of the population in the surrounding towns and villages. In fact, the employment rate is higher in these cities than the average of the whole island (37.3 percent versus 35.4 percent). The rural residuum is, on the other hand, rather poor. The farm sector is mainly composed of small farms located in hilly or mountainous areas where rural industrialization is almost nonexistent. In the rural areas of both Sicily and Calabria, welfare benefits and funds sent from relatives who have migrated abroad represent a large source of income for the local population (Cacciola, 1980). The per capita income of Sicily ranks higher than that of Calabria but is still 7 percent below the national average. The farm sector in both regions is characterized by a strong dualism between large and small farms. In Sicily there has been a sharp decrease in the last two decades in the labor force employed in the sector. In 1961 the population employed in agriculture was 610,000 (41.4 percent of the working population of Sicily); by 1971 it had decreased to 324,000 (27.3 percent), and by 1976 it was 285,000 (25.9 percent). This percentage is still quite high when compared to that of the country as a whole (15 percent) and to the North (9 percent). In 1961 there were 557,745 farms; by 1971 their number had decreased to 465,240, and in 1976 to 400,645. Of these farms 42 percent are less than two hectares in size.

A total of 189,215 hectares of land is occupied by forests, which constitute 7.8 percent of the total agrarian territory of Sicily (compared to 22.5 percent of all Italy).

The structure of the agricultural labor force in Sicily shows a sharp trend toward proletarianization (Furnari, 1980; Nocifora, 1980). The percentage of farmers in the period since 1969 dropped from 31.1 percent of the agricultural working population to 29 percent, while the percentage of farmworkers increased from 60.8 to 64.8 percent.

In spite of the tendency toward proletarianization of the labor force, reduction of the number of farms, and degradation of the land, Sicilian agriculture, as a whole, has exhibited a productive development in the last decades. Gross production increased drastically by a record 87 percent. This increase was larger than that achieved by more prosperous and advanced agricultural areas. Lombardy, which is the richest region of Italy, showed an increase of 76 percent, Veneto showed a 78 percent increase, and Emilia, which is the most advanced agricultural region in Italy, showed an increase of 74 percent. In Sicily the record increases are due to the expansion of two sectors: citrus fruits and vegetables (Progetto Sicilia, 1978). Increases were also achieved by the grape sector, while grain production and cattle and hog breeding displayed a decrease. A decreasing production was recorded in olive oil and nuts production as well, which were once two of the most common crops in Sicily. In other words, all of the increases that were recorded during the last decade may be attributed to the expansion of dynamic cultures on irrigated land (Nicifora, 1980), that is, on land that makes up only 7 percent of all the cultivated area in Sicily but produces 42 percent of all the agricultural commodities of the island. Furthermore, such a growth took place in spite of the EEC agricultural policy that penalizes the production of fruits and vegetables.

Agriculture in Sicily is characterized by a contradiction between a critical occupational situation and a growth in production. This contradictory situation has been related (Mingione, 1981) to the structure of Sicilian agriculture, where a large number of small and medium farms, especially in the internal areas, became increasingly marginal, and a group of advanced farms expanded greatly. Economically advanced farms occupying the most profitable land, are larger in size and display high levels of productivity (Mingione, 1981).

The conclusions reached for Sicilian agriculture can also be applied to Calabria. In that region there was a sharp reduction in the number of farms in the past decades from 465,879 in 1961 to 295,254 in 1980. Of the farms remaining, 40 percent are two hectares or less in size. Production rose in the fruits and vegetables sector, but this increase was not as large as that of Sicily. Grain, cattle, poultry, and hog production decreased sharply (ISTAT, 1980). Irrigated land represents 8 percent of the total land available in the region and is occupied by the best farms. The percentage of farmers among all agricultural workers decreased in the last ten years from 36.3 percent to 31.9 percent, while the percentage of farmworkers increased from 53.7 to 56.7 percent (Furnari, 1980).

The small farm sector as recorded in our survey presents the following general characteristics. From the structural point of view it is largely composed of farms that do not exceed 5 hectares (Table 4). As only 6 percent of all farmers interviewed cultivate a larger portion of land (Table 5). The most common cultures are fruits and vegetables. About 50 percent of all farmers surveyed grow only these crops, while almost all the others combine fruits and vegetables with other cultures and breeding activities (Table 6). Small farms are not divided into many parcels. 54.3 percent of all farms are formed of one parcel, while an additional 17.1 percent are formed of two. Only 28.6 percent are farms with more than two parcels, and among them only 16.5 percent have 4 or more (Table 7). The parcels are also located near the farm, for 80 percent of all parcels are within a range of three kilometers from the family house (Table 8).

The majority of small farms have annual sales of less than 8 million lira ($6,000), while only 16.5 percent have sales over 12 million lira (Table 9). The net income produced by these farms is rather low. 41.2 percent of them earn less than 1.5 million lira annually while 7.5 percent generated an income of over 10 million lira which is still below the average for all farms in the South (Table 10). Farming, however, is the sole economic activity for only 24.1 percent of all farm families (Table 14). All other families have at least one member engaged in off-farm activities. The relative unimportance of farming as a source of income is supported by the fact that over 60 percent of all families obtain more than 80 percent of their total income from off-farm activities, while another 14.1 percent earn between 60 to 80 percent of the total household income from off-farm jobs (Table 11). Moreover, families that have a total income of 10 million lira or less account for only 21.6 percent of the sample, while 45.7 percent have an income of over 20 million. Among them 6.5 percent have an income of over 28 million (Table 12). Additionally, 73.8 percent of farm families' members declare that the farm income per se would not be sufficient to support the family in any case. Another 9.6 percent state that the farm income is not sufficient for the family, but it could be sufficient if the standard of living were to be somehow lowered (Table 13).

Of all families 58.8 percent have at least one member with a full-time off-farm job (Table 14). Among the off-farm jobs held by family members, the majority is represented by clerical jobs, usually in administrative agencies such as local town halls, or in banks, stores, etc.. The individuals who hold these jobs tend to be young, and usually they are not heads of the households. Few people are engaged in agricultural off-farm activities (11 percent). The most common of the blue collar jobs are those in construction, and janitorial services (Table 15). In spite of the large amount of off-farm activities, farm work still absorbs a substantial part of the working day of family members. Among heads of the household, only 9.5 percent work twenty hours a week or less on the farm, while 43.7 percent work more than 50 hours. A bimodal distribution is noted as far as the farm work of other family members is concerned. They either work only part-time or above the regular full-time load (over 48 hours a week) (Table 16). A difference exists between male and female work. Females tend to work more as part-timers than males. This trend supports the conclusions reached by other recent studies of female participation in farm labor (Furnari, 1980, Pugliese, 1983). It has been stressed, in fact, that female contribution to farm activities is decreasing. It reached its peak during the middle seventies and then decreased due to the availability of male labor, generated by the stagnation of the labor market.

The abundance of male and female family labor does not impede the use of hired help on the farm. Over half of all farms surveyed use hired labor (52.8 percent). This labor cannot be substituted for family labor because family members either refuse or cannot perform the required farm activities. For those families who might substitute hired labor with family labor 87 percent cannot do so, because hired labor is largely used for special operations such as harvesting and pruning (95 percent) and is used in a limited amount. In fact, only 12 percent of all farms sampled hire labor for more than 250 days a year, while over 50 percent hire for only 100 days or less. The salary paid to farm workers tends to be lower than the one fixed by law, 40,000 lira a day (for 1983) or approximately $25. Only 10 percent of our sample pay such a salary, while 35 percent pay below half of the rightful salary to farm workers (Table 17). The availability of labor is also indicated by the low employment rate (the ratio between work-elegible population and employed population), which is below the level recorded for the whole South (31.2 percent versus 35.5 percent).

Their small size does not prevent over half of the farms in our sample from receiving financial aid from Italian and European agricultural agencies. The amount of this aid is rather limited, as it averages about 3 million lira per farm ($2,000), while the average for all farms in the South is 11 million. It usually takes the form of either payments for price support programs, particularly for olive oil production, or incentives for farm improvement such as irrigation or culture amelioration.

Investments are, in general, quite low; 57.8 percent of all farms surveyed did not make any investment in the past five years. The other 42 percent invested primarily in the area of culture improvement. Structural investments (construction of buildings or purchase of machinery) account for only 8 percent of the total amount among all farms in the South. According to nationally issued data (ISTAT 1983), over 80 percent made some investment during the previous five years. However, this percentage decreases to 65 if farms of 5 hectares or less are considered. The low level of investments is reflected in the limited presence of machines on the farms, for only 1 percent of all farmers invested in machinery in the last five years. In our survey 21.6 percent of all farms do not have any machinery, while 56.8 percent have only one or two tractors (Table 18). However, the limited average size of farms and the availability of labor have been traditionally used to explain the low level of mechanization. Such an explanation seems to fit our case also.

The land market in South Italy has been rather static in the past decades, as the high price of land and decreasing returns have discouraged transactions (De Benedictis, 1980). This trend is confirmed in our survey (62 percent of all farms were inherited, while only 21.1 percent were purchased). However, signs of a more dynamic market have been recorded in the last few years. If farms that have been acquired since 1974 are considered, a sharp increase in purchases is shown. They increased over 50 percent in this period from the previous decade. Inheritance, though, still represents the more common manner of obtaining a farm.

The vast majority of farms are owned by the cultivator (84.4 percent). Only 2 percent of all farms are rented, and the rest are either partially owned and partially rented (9 percent) or managed under a sharecropping contract (4.6 percent). Not all of the farms' products are marketed; a substantial portion (37.2 percent) use most of their products (90 percent or more) for family consumption. On the other hand, another 36.2 percent of farms are largely market oriented whereby at least 80 percent of their products are commercialized (Table 19). Previous analyses of the small farm sector in the South (Mingione, 1981) showed that family consumption of farm-produced goods is high throughout the region. The relative quantity of farm products grown and consumed by the farm family in our sample is also related to farm income. In families with higher farm income, consumption of farm goods tends to be low, while it is high on farms with low farm incomes.

The age distribution of the farm population in the sample is different in some regards from that of both Calabria and Sicily taken as a whole (Table 20). In the sample there are more young and old farm members, while the percentage of people in the central age groups is inferior to that recorded for the two administrative regions. Young farmers are rarely heads of households. In fact, only 4.5 percent of all the farm households have a head who is 40 years old or younger. Heads of households over 60 years of age account for 47.3 percent, and among them, 17.1 percent are aged 65 or more.

A final datum regards satisfaction with farm life. Heads of households tend to be more satisfied with farm life than other family members, for only 39.7 percent of those who were not heads of households said they were very satisfied. This percentage rises to 59.2 percent in the case of heads of households (Table 21).

In summary, it appears evident that the small farm sector and the geographical region surveyed are marginal in respect to all other sections of the country, the rest of the South and the agricultural sector as a whole. The marginality of this region provides the background for the main research question of the present work, the persistence of small farms. This issue is addressed in the next chapter through the illustration of some specific farms cases. Such farms are described by their structural characteristics and by the structural characteristics of family members. These data are dialectically combined with family members' accounts of the history of their farms. Structural data and personal accounts constitute the supportive evidence for the conclusions in the research.

5
The Persistence of Small Farms

IDENTIFICATION OF FARM GROUPS

The farm families interviewed have been divided into six groups on the basis of their primary motivations for remaining in agriculture. The concept of "motivation" refers to the rationale behind the decision of each farm family to remain in agriculture. Though this concept is extremely useful in analysing the phenomenon of persistence, it presents some problems which merit a brief discussion. First, motivations are based on a multiplicity of factors. In the typologies identified in this book, for instance, we find families whose motivations are dictated primarily by economic concerns, others by emotional concerns and still others by varying combinations of the two. Furthermore, the "motivations" behind persistence reflect the rationalizations of the farmers' life conditions which may appear different to each individual even in the case of similar structural circumstances. In this respect we may find families who experience similar structural situations, but whose interpretation diverges considerably. The latter consideration is connected with the "second" problem, the lack of clear boundaries among the groups considered. As the reader will notice, there is a great deal of overlapping among the groups so that families classified in one group could, with some reinterpretation, be classified in another. Reduction in the number of typologies has been discarded, however, because, despite the problem of overlapping, a great deal of information is preserved and important differences are maintained. Though the typologies are not sharply separated, each distinctly contributes to the understanding of the complex and contradictory patterns of the persistence of small farms.

The six groups of farm families are defined as follows:

- The first group, named "Traditional," is composed of families who keep their farms due to emotional attachment to the land and/or farm life.
- The second group, called "Reserve," contains families who do not have available job alternatives to farming or do not have skills that will allow them to leave the sector.
- The third group, classified "Complementary," is composed of families who remain in farming because the income from this activity complements that of family members with off-farm jobs.
- The fourth group, which is referred to as "Retired," consists of families in which one or more members have retired from previous occupations and invested their savings in the purchase of a farm or in a farm that they already owned but did not actively cultivate.
- The fifth group, that of "Equity," is also composed of families who have invested in the land, though their reasons for doing so differ from those of the previous group.
- The sixth and final group, which is a "Residual" group, con-sists of several subgroups of families who remain in farming for differing reasons. Among them, three assume particular theoretical relevance. The first is the subgroup of families who view farming as a hobby to be pursued during leisure time. The second consists of families who are ready to sell the farm and will do so at the first available and convenient opportunity. Finally, there are families who would like to sell the farm but consider the offers received not adequate for the real value of the farm.

Before presenting specific cases of farm families belonging to each group, it is important to draw a brief quantitative profile of each group. To this end, the differences between and among groups have been analyzed with the aid of the technique of analysis of variance. A significance level of .05 is employed, even though non-statistically significant relationships are also described. Although the use of the significance level is extraneous to the epistemological framework employed in this work, its use is motivated by the fact that it can provide a set of relevant indicators for interested readers. Moreover, it can also provide means for the comparison of this study with others on the same or related topics that employ quantitative techniques of data analysis.

Finally, it is relevant to mention briefly a typical characteristic of Southern Italian families. Within the Southern Italian family, offspring tend to remain within the family well beyond the age of 18 and in some cases even to the age of 30 (however, the family as such is not very large in size -see table 23-, containing an average of 4 individuals) (Barbagli, 1977). The reasons for such a phenomenon may be attributed to economic, structural, and emotional factors. A lack of jobs and/or jobs that pay well forces some young people to remain with their parents. Due to the low income of the parents, the economic support provided by these offspring is often needed by the family (Barbagli, 1977; Sgritta and Saporiti, 1980). In addition, it has been noted that the relatively high price of dwellings in both rural and urban areas also encourages cohabitation between parents and adult offspring (Ginatempo, 1975). Finally, a cultural component having to do with strong family ties has been indicated as an explanation for cohabitation (Ardigo and Donati, 1976; Saraceno, 1976).

THE GROUPS

Traditional Group

This group is composed of families that remain in agriculture due to an emotional attachment to the land and/or farm life. It represents the largest group, containing 38.7 percent of all families interviewed.

The average total income of these families is higher than the average of the sample (Table 28). However, it ranks only third in total income among the six groups considered in this study (Table 28). As far as the composition of total family income is concerned, it must be noted that it is largely of off-farm origin (an average of 58.7 percent, with 21 percent of the families obtaining 90 percent or more of their total income from off-farm sources), though this component is the next to the lowest among the six groups studied (Table 24).

These farm families also differ (this difference is statistically not significant) from the rest of the sample surveyed in their account of the reasons that motivate other families to abandon farming. In fact, although there is a great uniformity among groups, it is apparent that farms are abandoned primarily for economic reasons. 12.5% of the members of this group (the highest percentage among all groups) state that people leave farming due to their dissatisfaction with the style of life that it provides. Moreover, the majority of both heads of household and other family members in this group are satisfied with the farm and farm life (Table 26).

These farm families tend to be relatively younger than those in other groups, though the average head of the household is 59 years of age (Table 35). Finally, the size of farms in this group is below average (Table 27) while the amount of land utilized is above average (Table 36).

CASES

Four cases have been selected to illustrate the specific characteristics of this group of farm families.[9]

First Case

The first case involves a family of two, a 61 year old husband and a 58 year old wife, who live in the county of Benestare in the Northeastern part of the region surveyed in Calabria. The family was originally com-posed of four members, but the two daughters left the household after marriage. Nevertheless, they still contribute to the farm enterprise by working an average of 25 hours a week, primarily during the harvesting season. No hired labor[10] is employed on the farm.

The husband supplements his work on the farm with jobs in construction, a sector that served as his major occupation for over two decades from the mid-fifties to the mid-seventies. He also worked part-time in agriculture as a hired worker until only a few years ago. At present he is involved in no activity as a hired worker, and his job in construction has been reduced to a minimum due to his age and the limited labor demand existing in the sector. He spends most of his working time on the farm, where he works over 60 hours a week. In addition to the regular compensation for these jobs, he receives several types of welfare support. One of these is unemployment benefits for the periods in which he is not engaged in construction work. Simultaneously, he receives support as an unemployed farmworker. Both pensions are extended to all periods of unemployment without time limitations.[11] these benefits are rarely revoked. One example is provided by the unemployment pension for farmworkers. Periodically, a list of all the unemploymed farmworkers in the country is compiled. All of these workers are entitled to receive unemployment benefits in accordance with the number of days worked during the previous year. Such payment is continued until unemployment ceases, and it is still maintained if the individual finds employment outside of agriculture. As Mingione (1981) suggests, these individuals receive a tenured pension.

The wife does not have a regular job but works periodically as a farmworker on a neighboring farm. This activity is usually performed during the harvesting seasons for citrus fruits (January-March) and for grapes (September-October). For the rest of the year, most of her working time (40 hours a week on the average) is spent on the family farm. She also receives welfare payments as an unemployed farmworker.

The farm was inherited in 1950 when the head of the household's father died. Since then the family has not sold any part of it nor purchased additional land, so the size of the farm is the original seven hectares (15.2 acres), all of which are currently utilized for agricultural purposes. 60 percent of the land is a vineyard, while the remaining 40 percent is divided between citrus fruits (70 percent), vegetables (20 percent) and grains (10 percent). Five milk cows are also present of the farm.

The farm receives economic aid from the EEC in the form of price support payments for the production of olive oil, grain, and dairy products. The total net farm income is 10 million lira, including the contributions of the EEC. Family members state that about 65 percent of their family income is off-farm origin and that 40 percent of all farm products are not sold but consumed by the enlarged family.

The head of the household, as well as his spouse, enjoy farm life. They also stress that the income generated by the farm alone would be sufficient to support the family, though a reduction in their standard of living and an increase in economic aid from the State would be necessary. However, they doubt that the farm would ever be the only source of income for them. Their stated commitment to farm life and the land, in spite of insufficient economic gain, is reinforced by their account of their reasons for staying in farming. Says the husband: "I inherited this farm and worked here first with my parents and then with my family. I don't want to leave it because it gives me a sense of independence that nothing else can give me, and it helps me continue a family tradition." Similar feelings underlye their understanding of the reasons why other farm families leave. "Farming is hard work," the head of the household declares. "People who leave do not like to work hard and don't love the land, especially the young. If you really love the land, you stay."

This explanation, however, contradicts the structural data collected for the farm, for hard work does not seem to be the key to its success. This farm generates less than half of the total family income and could hardly guarantee by itself a sufficient socio-economic standard of living for the family. Moreover, farm work has traditionally been supplemented with off-farm jobs. Given the family's commitment to farm life, this dependency upon alternative sources of income testifies to the objective need for substantial additional economic support.

Second Case

The second case involves a family of two, husband and wife, from Stignano county in the Eastern portion of Calabria. They are both in their late fifties (59 and 57) and have a limited formal education (fifth grade). Each works on the farm for an average of 30 hours a week with no other family members or hired labor employed. They also work off-farm as proprietors of a small grocery store in the village of Stignano. In addition, the wife receives a pension as a retired farmworker. From the store and the pension, they receive a net off-farm income of about 15 million lira, which constitutes 75 percent of their total income.

The farm is quite small, consisting of only two hectares divided into two parcels of one hectare each. It was acquired in 1956, half through inheritance and half through purchase. Its major cultures are fruit, olive trees, and citrus fruits, which occupy almost equal portions of the land. The percentage of products sold is only 20 percent of the total, providing a net income of 3 million lira, while the remainder of the products are retained for consumption by the family. The farm enterprise also receives funds from the EEC price support program for olive oil.

The members of this family are not satisfied with the income generated by the farm, attributing its low returns to both its small size and to the mediocre fertility of the soil. In addition, they state that farm income is not sufficient to support the family and that it probably never will be in the future. This insufficient income, they stress, motivated them to initiate off-farm activity, which is now their major gainful occupation.

The same pessimistic attitude is apparent in their account of the reasons that motivate neighboring families to abandon agriculture. It is their opinion that in the last ten years about 50 percent of all farm families have left the area. This outmigration is due to both the low standard of living provided by agricultural activities and a lack of off-farm opportunities that might enable them to remain on the land as part-timers. Consequently, the family members consider themselves privileged to be able to combine farm work with an off-farm activity. They love the land and this attachment, they say, is their only reason for remaining on the farm.

The family members' understanding of the concept of "love toward the land" clarifies their attitudes in relation to a number of social values. This understanding is also common to almost all other family members in this group. For them love toward the land has a very precise meaning that can be summarized as a strong sense of family membership and an appreciation of traditional values and the way of life typical of farming. Such an understanding is not very different from that of farmers in the United States (agrarianism) and in other Western countries (Flinn and Johnson, 1974; Mingione, 1981). In fact, the way of life they like to identify with farming centers around values such as hard work, frugality, honesty, and the superiority of farm work over other activities, though they acknowledge the higher standard of living offered by urban life. In this respect it appears that in some instances their attachment to the land is a reaction to their inability to obtain a standard of living equal to or better than urban ones. The husband states, "We stay on the farm because we belong here. Perhaps, this is not a nicer life than that of the city or of the North, but this is what we have and what we want to keep."

Once retired from the family's town business, husband and wife would like to make agriculture their full-time activity. They are fully aware, however, that the possibility of retiring on the farm is related to the continuous availability of an off-farm income.

Third Case

This case involves a family of four, a husband and wife of 33 and 29 years of age, respectively, and two children of 5 and 3. They reside in Pezzolo county, which is located a few miles south of the city of Messina in the Northeastern part of Sicily. Of the two adult family members, only the husband works regularly on the farm, averaging 25 hours a week year round. He is assisted in his farm work by two other members of the enlarged family, his parents, who perform farm activities for an average of ten hours a week. Furthermore, additional labor is employed primarily during the harvesting seasons. For 1982 a total of 20 working days were paid to hired farmworkers.

Both husband and wife have a college degree and are employed full time off-farm. The husband works for a government agency, while she has a job in a county office. Due to the amount of their combined off-farm income, which total over thirty million lira, they do not receive any welfare payments.

The farm was purchased in 1980 and consists of a total of two hectares. Of these one and one half hectares are employed for the cultivation of oranges, while the remainder of the land is used for rabbit breeding. The production of rabbits is very popular among Italian farmers, as this meat is a staple in the diet of a substantial portion of the country's population.

Although the orange trees were already productive when the farm was purchased, investments were made during the past two years for the improvement of their condition. Additional investments were made for the construction of facilities for rabbit breeding, which was initiated after the purchase of the farm. The owner dedicates most of his farm working hours to this enterprise, while his parents and hired labor concentrate their efforts of the production of citrus fruits.

With regard to his reasons for purchasing the farm, the head of the household states that his motivations were related to his farm background. Born and raised on a farm, he felt that the purchase of his own represented a return to his origins. He also expresses a desire to expand his farm activity so that in the near future he can leave his town job and dedicate himself to farming full time. In stressing his commitment to farming, he indicates that only a bankruptcy of the farm enterprise would motivate him to leave.

This enthusiasm for farming remains even though the farm enterprise has yielded rather low returns in the last two years of operation. For the 1982 agricultural year the net farm income did not exceed two million lira or 5 percent of the total family income.

As far as his reasons for remaining in agriculture are concerned, the owner of the farm indicates that love for the land is his primary motivation for staying. He adds, however, that the sector is composed of many farms that, though limited in their productive range by the small amount of land available, are not rationally organized. He wishes to prove with his work that a rational organization of farm production and the "right kind of products" can eventually generate a sufficient income for the farm family even in the case of a small size enterprise. For the time being, he admits, his farm is far from reaching this goal.

The family's account of the reasons that motivate other farmers to leave the sector derive from this understanding of the situation in agriculture. According to both husband and wife, the people who left agriculture in the area were largely dissatisfied with it. Young people in particular prefer a more secure type of employment that provides a more adequate income. Moreover, the hardship of farmwork is less tolerated by the new generation. In other words, they conclude, "you must love the land in order to endure the sacrifices and hard work that farming involves."

Fourth Case

The last case examined in this group involves a family of three living in the Southeastern part of Sicily twenty miles north of the city of Catania. This family is composed of a husband and wife, 40 and 34 years of age, respectively, and a son of 15. The husband, who is the only family member with an off-farm job, works as a clerk for the State-owned railway company in the nearby town of Giardini. He also spends an average of 20 hours a week working on the farm. His wife and son also contribute to farm activities. She works for 10 hours a week, while the son provides assistance during his time from schoolwork. Two or three hired laborers are also used for a period of three to four weeks during the lemon harvesting season.

The farm, which is consolidated in one parcel, was inherited in 1971 from the head of the household's father, who retired. It consists of five hectares which are primarily devoted to the production of lemons, though vegetables are also grown in minimal amounts.

In recent years very few investments were made in the farm due to the unstable condition of the regional lemon market, which showed a steady decline in price. Consequently, the owner of this farm did not expand production and limited farm investments to a minimum. Most of the production, however, is commercialized. All of the lemons are sold to a middleman, producing a net income of 2 million lira, which represents approximately 15 percent of total family income. The vegetables are, on the other hand, consumed entirely by the family. No welfare benefits are regularly received by family members, and no financial aid is granted to the farm by domestic or international organizations.

The family members are not satisfied with the farm primarily because of the income it generates, which is too low to provide an adequate stimulus to endure the hardships of farm life. In fact, they have considered leaving the farm on a number of occasions but have not yet done so. The reasons for this ambivalence are explained by the head of the household in terms of family values: "I began this activity when I inherited the farm from my father who received it from his father. If I were to leave it or, even worse, abandon it, my parents would be very disappointed, and I don't want to displease them. We stay on the farm because it represents a value to them, and we want to share the same values." The factors that have motivated this family to remain on the farm are thus entirely emotional. Yet, these feelings are not directed toward the farm or farm life itself but rather toward the enlarged family who has lived on the farm for generations. It is a sense of family membership and respect for the elder members of the family that overcomes in this case the clear dissatisfaction with the farm enterprise and the income it generates.

Traditional Group Summary

The principal characteristic of this group of farm families is that they remain in agriculture due to an emotional attachment to the land and/or to farm life. They also display strong traditional values which recall those considered characteristic of "agrarianism" in the United States. All of them, however, rely heavily on off-farm income to support the family, and without such income their farm activity would be almost impossible. Nevertheless, the awareness of such a dependency is not homogeneous among them. Some farm families, as in the first case presented, believe that they could obtain a sufficient income from the farm enterprise alone. Others, as in the third case examined, feel that this may be possible in the near future. The other two cases involve families who do not make any claim of obtaining an adequate farm income. However, in both the first and third cases the possibility of a sufficient farm income is purely hypothetical. In the first case there is no concrete evidence to support the future realization of this goal, while the third case can be verified only in the future.

Reserve Group

The second group is composed of farm families whose members do not have job alternatives outside of farming and/or lack skills that would allow them to have a sufficiently remunerative job outside the sector. Consequently, they are forced to continue farming because it represents an extremely important, if not principal, source of income. They also would leave farming if they could. In this respect their position in the labor market is that of "industrial reserve army" ready to be employed elsewhere when the occasion arises. These families account for 20.1 percent of the sample surveyed.

From an economic point of view the principal characteristics of this group can be summarized as follows. First, its members have the lowest family income among the six groups analyzed (Table 28). Second, their net farm income is the second highest among all groups studied and is well above the average of the sample (Table 29). Finally, this group recorded the lowest percentage of total income of off-farm origin (41.6 percent) (Table 24), which underscores the relevance of farm income for the persistence in agriculture of these farms.

This group contains over one-third of all farm families who do not have a member who holds an off-farm job (15 families out of 45) (Table 30A). Additionally, the proportion of off farm part-time jobs is higher here than that recorded for the rest of the families surveyed (Table 31), thus indicating the precarious nature of off-farm employment in this group.

Compared with the remainder of the sample, farms in the Reserve group tend to be smaller in size, ranking fifth among the six groups in this category (Table 27). However, the amount of land utilized on these farms is above the average of the groups (Table 36). A larger percentage of them have been inherited by their present owners, while a smaller proportion have been constituted by land reform action (1952-1960) or inherited and purchased (Table 32). Finally, the amount of hired labor employed does not differ from that used in other groups. However, the wages paid to labor are lower than those of the rest of the farms (Table 33).

CASES

The characteristics of the Reserve group are illustrated by means of four different cases.

First Case

The first case of the reserve group involves a family of two members residing in Pezzolo county in the Northwestern part of Sicily. They are a husband and wife, 60 and 57 years of age, respectively.

Both work on the farm for an average of 30 hours a week. Additional help is provided by their sons, who work for an average of 10 hours a week with peaks during the harvesting season, and by hired labor who are employed for approximately 10 days during the harvesting season. The husband spends a considerable portion of his working time as a farm laborer on other farms in the area, totaling 200 working days in this capacity. He and his wife also receive welfare payments as a result of both their indigence and their status as unemployed farmworkers. The income provided by these benefits and off-farm activities constitutes 65 percent of the total family income.

The farm, which was purchased in 1950, is 5 hectares in size and is divided in three parcels, all located within a range of 2 miles. It is cultivated primarily with citrus fruit and vegetables. There are not major pieces of machinery other than a tractor, and no investments have been made in the past few years. All of the vegetables and 5 percent of the citrus fruit production are consumed by members of the enlarged family. Only the remaining citrus fruits are commercialized, an activity that generates a rather low net farm income estimated at 3 million lira.

According to the members of this family, several farm families have left the area in the last ten years. The owner attributes this phenomenon to the unprofitable nature of the citrus fruit culture. The saturation of the market, which pushed prices down, coupled with the limited production of these farms generated returns inadequate to even cover the costs of production. This difficult economic situation is also identified by family members as a major cause for overall dissatisfaction with farm life. Both husband and wife declare they they have considered leaving farming on more than one occasion. However, three principal considerations have prevented them from doing so. First, there are no other jobs available for men or women of their age that would provide a more adequate income. Second, as the farm and the farmhouse are not very valuable on the market, their sale would constitute an economic loss. In the case of the house, the high market prices would make it impossible for them to replace their present dwelling with another of comparable quality. Finally, the farm enterprise provides an alternative to working under the supervision of someone else. It allows them, at least partially, the freedom of managing their own business.

Second Case

The second case in this group involves a family of five people living in Montestarace county in the Southeastern part of Calabria. The family is composed of a husband and wife, two daughters, and a son. The husband is 57 years of age, the wife 55, the two daughters 25 and 26, and the son is 15 years of age.

Only three family members perform regular work on the farm. The two parents each work for an average of 65 hours a week, while the eldest daughter contributes an average of 15 hours per week. Three laborers are also hired each year for a total of 20 working days to help with irrigation and pruning operations. They are paid 35,000 lira a day and "some wine."

Of all the family members, only the second daughter has an off-farm job, working as a substitute teacher in the county high school. The husband and wife receive welfare payments primarily in the form of unemployment benefits. These payments total 5 million lira yearly, while the daughter's wages for the past year barely exceed 3 million lira. These two income sources account for over 40 percent of the total family income.

The farm was purchased in 1957 and is three hectares in size, all of which are consolidated in one parcel. Though the principal culture is that of citrus fruits (70 percent of all land), olive trees (20 percent of land) and other fruit trees (10 percent) are also grown. In the last five years there has been one major investment, the replacement of a tractor.

Most of the agricultural goods produced on the farm are commercialized and are sold directly to the public by the family, which has a stand in the local open market twice a week. A total of 10 percent of the farm production is consumed by the family itself. In addition to the returns from this commercial activity, the farm receives price support payments from the EEC for olive oil production, which helps generate an annual net farm income of 10 million lira. Family members consider this income insufficient by itself to support the farm family.

The family's attitudes toward farm life are different, depending upon whether the head of the household or the rest of the family members are considered. The husband states that he is satisfied with farm life, but this feeling is qualified somewhat by the low farm income. On the other hand, the other family members are clearly dissatisfied with the farm and its lifestyle and desire a change.

Dissatisfaction with farm life and the lack of an adequate income are also the principal reasons for farm outmigration, according to family members. They state that these two factors have motivated several local families to abandon farming during the past decade. The husband notes that the idea of leaving the farm has also been considered within the family, though such an action has never been taken. The family members, accounts of their reasons for remaining on the farm are quite contradictory. When asked to explain their motivation, they stress their attachment to the farm and their feeling that it would be a shame to leave it. However, when more specific questions were asked, they clearly state that they became and remain farmers due to a lack of other job opportunities and due to their lack of professional skills and education. Moreover, they emphasize that if a more remunerative job opportunity were offered them, they would abandon farming as their principal activity and take the new job. Additionally, it was made explicit that neither of the children would continue farming in the future. This last consideration seems to indicate that the lack of an employment alternative is the family's principal reason for remaining on the farm. The emotional motivations first mentioned (which were contradicted by farm family members in their accounts of farm life satisfaction) appear, then, to legitimate a situation that is largely outside the control of the participants.

Third Case

This case involves a family of four living in Torregrotta county in the Northwestern part of Sicily. The farm family is composed of a husband and wife of 60 and 53 years respectively, a 30 year old son, and a 27 year old daughter.

Of the four family members, only the parents work regularly on the farm, averaging a weekly total of 40 years for the husband and 35 for the wife. As the son's labor is limited to weekends, holidays, and vacations, his overall contribution is quite small. Nevertheless, this minimal assistance makes the employment of hired labor unnecessary.

The head of the household and his spouse do not have off-farm jobs but receive off-farm income generated by welfare payments, which totaled 2 million lira for 1982. The son and the daughter, on the contrary, do have full-time jobs. The son works as a clerk in the local town hall, while his sister has a job in the post office. Though both are single and enjoy economic independence, they do not yet wish to leave the household. Their economic contribution to the family budget is of great importance, as a substantial portion of their incomes is employed to cover most of the household expenses. The total net off-farm income is 20 million lira.

The farm is six hectares in size and is sharecropped. The original sharecropping contract was signed in 1954 and has been continually renewed since that time. However, in 1978 the Italian ligislature passed an act that prohibited the renewal of sharecropping contracts and called for a dissolution of all contracts in effect within ten years from the date in which the act was passed. The farm's contract expires in 1987, and the two parties have mutually agreed that the tenant family will have the first option for purchase.

Of the six hectares that form the farm, four are operated for the cultivation of oranges, one and one-half for the cultivation of other fruit trees, and the remaining land is used for vegetables. The vegetables production is mostly consumed by the family, while the oranges and other fruit are sold to a middleman.

There has been a limited amount of investment in the farm during the past five years, which involved the improvement of the orange trees and the purchase of irrigation equipment. These investments were shared by the tenant and landowner, 75 percent and 25 percent, respectively. It was agreed, however, that the percentage for the improvement of the orange plantation would be discounted from the selling price of the farm in the event of purchase by the tenant.

The farm enterprise receives financial aid from both the Italian Ministry of Agriculture and the EEC. The first agency provides financial assistance to this farm through its program of "Economic Intervention for the Development of Agriculture," which is implemented through regional administrative offices. Aid from the EEC is received through the European program for the amelioration of farms and, specifically, from funds designated for the improvement of irrigation facilities. The net farm income for the year 1982 was estimated at 9 million lira.

The family member's feelings with respect to farm life are not homogeneous. While the parents enjoy their lifestyle and express a strong satisfaction with their activity as farmers, the son and daughter would like to leave the farm. Their relative lack of involvement in the enterprise makes them less sensitive to feelings of attachment to the land. However, they state that their present income is not high enough to allow them to purchase a house that would satisfy their needs.[12] They have, thus, decided to remain in their parent's house until they have the opportunity to carry out their plans.

Structural concerns have also influenced the husband in his initiation and continuation of the farming activity. He recalls that when he first signed the sharecropping contract, farming was the only economic option available to him. Furthermore, he never mastered a skill that would allow him the possibility of other jobs. "Today," he states, "the situation has not changed. Farming still represents for me the only source of income, and it probably will in the future due to the high level of unemployment in this region." The unstable economic situation in farming is acknowledged by both husband and wife, who realize that without the financial help of their son and daughter, the farm enterprise would not survive. Nevertheless, they seem optimistic about the future.

"The farm," they stress, "even if not very profitable, represents for us the only possibility for the future, and we are convinced that we will never leave it."

Fourth Case

The fourth case involves a family of six living in the Siderno area in the extreme Northeastern part of Calabria. The family is composed of a husband and wife, 56 and 40 years of age respectively, and four sons whose age range from 14 to 3.

The husband and wife are the only family members who work on the farm. He works for a weekly average of 60 hours, while she contributes only 15 hours. Though no hired labor is employed, several relatives provide assistance during the harvesting season without any monetary compensation.[13] The wife also has an off-farm job as a laborer on other local farms, working a total of 101 days last year and receiving unemployment benefits for the period in which she did not work. Moreover, for the past three years the family received an annual cash bonus given by the State for the birth of their last son. The total amount of off-farm income for this family is 3 million lira, which represents 23 percent of the total family income.

The farm is two hectares in size but is fragmented into four parcels, which are scattered over a radius of four miles from the house. Inherited in 1965, it is devoted to the production of four crops. Olive trees represent the most extensive crop, while citrus fruits, other fruit trees (peaches and apples), and vegetables are also grown. All of the crops are primarily commercialized with only 20 percent being used for family consumption. This commercialization is accomplished through the direct sale of agricultural products to the local village market where the family rents a stand. The farm enterprise also receives financial aid through the EEC in the form of price support payments for olive oil production.

There have been a few investments in the farm in past years, and these were related to the improvement of the citrus fruits and the olive trees. In addition, some machinery for the production of oil was purchased. However, the head of the household states that major investments will be executed in the future, for it is his intention to purchase more land in order to expand the farm. He adds that land in the area is quite fertile, and with a few more hectares he hopes to bring the farm income to a level adequate to support the family. At present the net farm income does not exceed 10 million lira.

The family members are not satisfied with the farm and farm life. The husband states: "I am not satisfied with my activity as a farmer because the farm is not profitable. It is too small for an adequate use of machinery, and I would need more capital to expand. This capital is difficult to obtain because the State doesn't help farmers, particularly here in the South, as it should. We farmers need more from it." He also declares that in the past he left agriculture for a while and found employment as a clerk in a local business. When this employment was terminated, he had no alternative but to return to his farming activity full time. The reasons for his search for an off-farm job had to do with the fact that family members considered the farm income insufficient to fulfill the family's needs. Moreover, if no improvement in the economic situation of the family is recorded in the near future, all family members agree that a drop in the already low standard of living will occur.

With respect to the reasons that have motivated other area families to abandon farming in past years, it is the family members' belief that only a few families have left. Those that did so migrated to urban areas where there were greater possibilities for remunerative jobs, careers and more satisfactory lifestyles. Those who have remained are either poor or have an off-farm job that helps them economically.

The family members' understanding of the condition of farming in the area ultimately explains the contradiction noted in the interviews between the desire to make further investments in the farm and the lack of farm profits. The absence of job alternatives outside farming is the major factor that has motivated the persistence of the family in the sector. Consequently, it is within farming that a possible solution to overcome the family's precarious economic condition needs to be found. In the understanding of the family members, more investments and an expansion of farm production represents such a solution. Simultaneously, the structural difficulties associated with the attempt to expand farm income justify personal dissatisfaction with respect to the farm enterprise and farm life.

Reserve Group Summary

The principal reasons for these families' persistence in agriculture lie in a lack of job alternatives outside the sector and/or a lack of skills that would allow one or more family members to obtain a sufficiently remunerative off-farm job. The farm income becomes, then, extremely important for these families, as it constitutes the principal component (and in several cases the only component) of the total family income.

Nevertheless, the possibility for these families to remain in agriculture is also linked to the considerable amount of welfare benefits paid to family members as unemployed workers, usually as unemployed farmworkers, and to the existence of the off-farm income of one or more family members.

The off-farm income of these families is largely generated by people occupying a marginal position in the labor market. Their lack of skills (most of them do not have more than eight years of formal education), their relative old age, and the relative saturation of the labor market itself, do not allow them many alternatives. In this respect the farm enterprise represents a vehicle through which this labor force can be engaged in some kind of occupation and can continue its existence without creating major social tensions. In view of this situation, State intervention in the form of welfare payments assumes great importance.

Complementary Group

The third group considered in this study is composed of farm families that remain in agriculture because the income from this activity complements that of family members with off-farm jobs. In other words, these families are composed of part-time farmers whose principal employment is off-farm. The group contains 12.6 percent of all farm families surveyed and is the fourth largest group among the six considered.

From a demographic point of view, these families are composed of relatively young individuals. The average age of heads of households is, in fact, the lowest among all the groups (58 years old) as is that of the spouses (39 years).

From an economic point of view, it must be noted that this group of families has the highest average total family income among all groups (Table 28). Off-farm income as a percentage of total family income is also higher than that of the average of the sample (Table 24).

Though the average farm size is larger than that recorded for the rest of the sample (Table 27) there are not significant differences in the types of crops, use of labor, investments made on the farm, or total land utilized from those recorded in other groups of farms.

An important characteristic of this group is the family's attitude toward farm income, which is largely considered to be insufficient to support the family (Table 34). Moreover, economic factors are indicated as the only motivation that leads neighboring farm families to leave agriculture (Table 25). Finally, the percentage of family members in this group who are satisfied with farm life and the farm is the second lowest among all groups (Table 26). Conversely, the percentage of heads of households satisfied with farm life is the second highest (Table 26).

CASES

This group of farm families is illustrated with the use of three cases.

First Case

The first case of the Complementary group involves a family from Barcellona county in the Northwestern portion of Sicily. This family is composed of five members, a husband, wife, two sons, and a daughter. The parents are 57 and 55 years of age, while the two sons and daughter are 26, 22, and 18 years of age respectively.

The husband is the only family member working on the farm for an average of 15 hours a week. A limited amount of hired labor is also employed for a total of 30 working days annually. Three family members have off-farm jobs. The head of the household is a clerk in the local branch of the National Electric Company (ENEL); the first son, who has a high school diploma, works as a clerk in the town post office; and the second son, who has several years of college, works as a clerk in a construction materials factory. The three jobs together provide the family with an off-farm income of 40 million lira, which constitutes approximateley 90 percent of the total family income.

The farm is two hectares in size and was inherited by the family in the early 1970s. Consolidated in one parcel, it is dedicated to the cultivation of citrus fruits and vegetables. There have been a few investments in the last five years, notably, the improvement of citrus fruit trees and the purchase of a tractor. No plans for substantial future investments have been made.

Despite its limited size, the land is, according to family members, rather productive, and a net farm income of four million lira was made in 1982. This income was derived exclusively from commercialization of the citrus fruit production, which was sold on the market through the mediation of a middleman. The vegetable production, on the other hand, was used for family consumption. All family members indicate that the farm income is too low to support sufficiently the family and that it will not be the primary source of income for the family in the future either.

Family members' attitudes toward farm life are rather positive. Both parents and offspring declare that they enjoy life on the farm, even though the sons and daughters are certain to leave the household in the very near future.

According to the accounts of family members, two principal factors motivate their persistence in farming. The head of the household states that farming provides an economic supplement to the off-farm incomes. This complementary economic activity will be increased in the future when the off-farm income of the family is limited to that of the husband. Moreover, it would be an economic loss not to exploit the farm enterprise. "With moderate annual production costs," he continues, "it is possible to obtain quite satisfactory returns from this farm."

Second Case

This farm family is composed of four members, a husband, wife, and two sons (ages 50, 44, 21, and 19 years respectively) who live in Milazzo county in the Northwestern portion of Sicily. Of the four family members, three work regularly on the farm. The husband and wife spend an average of 15 hours weekly, while the youngest son contributes approximately 10 hours a week with peaks during the summer. Three to four workers are also hired for a period of one week to ten days every fall for the harvesting of olives. The other son does not work on the farm, as he has a full time off-farm job, which leaves him no time for other working activities. The head of the household has an off-farm job as well, serving as a clerk in a local administrative agency. These jobs generate a total off-farm income of 17 million lira, which is about 80 percent of the total family income.

The farm, which was inherited in 1971, consists of three hectares consolidated in one parcel. Its principal crops are olives, some of which are converted into olive oil on the farm, vegetables, and citrus fruits. Almost all of the products are sold, generating a net farm income of 3 million lira, while only 10 percent of this production is consumed by the enlarged family. The farm enterprise has been characterized in the past by a low level of investments, a policy that is expected to continue in the immediate future.

The family members' attitudes toward farm life and the farm are generally positive. Both the head of the household and the other members of the family declare themselves to be satisfied with farm life. The parents state that they have never considered leaving the farm, while the sons, though presently satisfied with farm life, plan to leave it in the future. As far as the family's motivation for remaining on the farm is concerned, they indicate that, in addition to the satisfaction it provides, the farm makes an important contribution to the total family income. The head of the household stresses that though the farm income is rather low, it represents an economic advantage in both monetary and non-monetary terms, which is difficult to ignore. "Moreover," he continues, "the labor required on the farm is not very demanding so that it is more a relaxation than a real job." Both from an economic and non-economic point of view, keeping the farm is considered advantageous by these family members.

Third Case

The third and final case of this group involves a farm family from St. Stefano county near the city of Messina in Sicily. The family is composed of four members, a husband and wife of 50 and 48 years of age, respectively, a 25 year old son and a 22 year old daughter.

Among the family members only the husband and son work on the farm, each averaging 20 hours a week. They are assisted in the cultivation of the lemon crop by hired labor, employed for a total of twenty working days per year. All of the family members have an off-farm activity, as they own a butcher shop in the nearby village of St. Stefano. This job is considered full time by all family members and provides them with 85 percent of the total family income.

The farm, which was inherited in 1960, is composed of six hectares consolidated in one parcel. The principal product is citrus fruits, mostly lemons, which occupy about 80 percent of the farmland. The remaining 20 percent is reserved for vegetables. There are also approximately 200 chickens on the farm, which are used for the production of eggs and meat and are sold directly to the public in the butcher shop. This commercial activity is one of the major assets of the shop, because it provides a direct and thus cheaper supply of farm products, allowing the store to be known as an outlet for genuine products directly from the farm.

The major investments made in the farm during the past five years involve the construction of chicken coops and the purchase of some pieces of machinery, including a new tractor. Of all the farm products, with the exception of the poultry production, 80 percent are commercialized through a wholesaler who acquires the various crops each year. The net income derived from this commercial operation does not exceed 6 million lira annually. Of the poultry production 95 percent is sold at the butcher shop.

All family members are satisfied with their farm lifestyle and labor. They not only view the farm as an important portion of their overall economic activity, but they also enjoy the living environment provided by the farm. It must be noted, however, that the low market value of land has further motivated the family not to sell the farm. The head of the household indicates that the value derived from its use is, in fact, much higher than the selling price available on the market. He also states that the farm income by itself would be insufficient to support the entire family. For this reason several neighbors who did not have any off-farm income to complement their farm income sold the land and left.

Complementary Group Summary

These farm families remain in agriculture primarily because they consider farming an important economic complement of their principal off-farm activity. As illustrated in the three cases mentioned above, the bulk of their income as well as the majority of the family's working hours come from non-farm activities. Nevertheless, a great deal of attention is lent the farm enterprise, which is considered by most family members to be a fundamental asset for the economic well being of the family itself.

At the same time, the farm is not viewed as solely an employment enterprise. These families receive a true satisfaction from living on the farm and participating in farm life. The agricultural labor required is very limited, and for this reason it is considered more a way of spending leisure time in a productive manner than an occupation.

Retired Group

The fourth group of farm families consists of people who declare themselves to be in agriculture because they have made farming their principal activity after retiring from another occupation. This group consists of 8.5 percent of all the families interviewed and will be referred to as "Retired."

The group's primary feature is the average age of its members, which is the highest in the sample (Table 35).

From an economic point of view, this group of families has the second lowest total income among all groups considered in the study (Table 28), however, its percentage of income of off-farm origin is the second highest among the six groups and averages over 80 percent of the total family income (Table 24). The majority of this income is generated by retirement pensions, rather than by current employment, a statistic reflected in the higher proportion of individuals in this fourth group with no off-farm jobs (Table 30A). Moreover, when off-farm employment is recorded by the family, it is a prerogative of offspring rather than parents (Table 30B, C).

As far as the farm enterprises themselves are concerned, most have been purchased or acquired in recent years (the year 1972 is the average). Owned for the most part of the cultivator, they tend to be smaller in size than those of the rest of the sample (7.00 ha and 7.74 ha, respectively).

Heads of household in this group are more satisfied with farm life than their counterparts in the other groups surveyed (Table 26). Moreover, all members of this group feel that families who have left farming have done so only because farm income was insufficient to support the family rather than for reasons related to dissatisfaction with farm life or farm work (Table 25).

CASES

Three cases are used to illustrate this group.

First Case

The first case involves a family of two, husband and wife, living in Gioiosa Jonica county in the Northern part of Calabria. They are 64 and 58 years of age, respectively and work exclusively on the farm, averaging 50 hours per week. Hired labor is employed for a total of 100 working days, mostly in the fall for grape harvesting.

The husband is retired and receives a pension. He and his wife migrated to the United States from Calabria over 40 years ago, both acquiring American citizenship while residing there. While in the United States he worked as a foreman, electing after retirement to return to his native country where he now receives a United States pension. As the wife never worked in the United States, she receives no pension. The total amount of this family's off-farm income is over $5,000 United States dollars a year.

The farm was purchased in 1981 when the family returned from the US. It is two and one-half hectares in size and is divided in two parcels of over one hectare each. Of the two parcels that, according to the husband, are quite fertile, the first is used for the cultivation of wine grapes, while the second is dedicated to the production of olive trees and vegetables. Most of these products are commercialized through a middleman with only 10 percent of them retained for family consumption.

There have been several investments since the farm was acquired two years ago. The most significant involves the improvement of the vineyard and the construction of a system of irrigation for the land devoted to vegetables. These investments were such that the returns provided by the farm in 1982 were barely sufficient to redeem them. Consequently, no disposable family income was generated. However, both husband and wife believe that the farm will generate enough profit in the future to support the family regardless of the off-farm income.

The family's attitude toward the farm and farm life is very positive. After the husband's retirement, the couple's first wish was to purchase a house in Italy where they would spend the rest of their lives. Their decision to purchase a farm, as well, was motivated by the fact that they grew up on a farm and wanted very much to return to that kind of lifestyle. Both family members stress that the farm provides them with an active yet relaxing activity to fill their retirement years. They have no intention of leaving the farm and plan to remain in agriculture for the next several years.

Second Case

The second case involves a Sicilian family residing in Gesso county, which is located Northwest of the city of

Messina. This family is composed of two members, a husband and wife, 63 and 55 years of age, respectively.

Both family members work on the farm, the husband for an average of 30 hours weekly and the wife for an average of 15 hours. In addition, 15 hours a week are contributed by one of the couple's sons who lives and works in the nearby town of Milazzo. Hired labor is also employed for a total of 20 working days, mostly in the spring for the harvesting of lemons. The wage paid for this labor (35,000 lira) falls well below the minimum level established by law.

Though neither family member has an off-farm job, the husband received a pension from the job he held before retiring in 1976. This pension pays a total of 12 million lira annually, which represents almost 75 percent of the total family income.

The farm was acquired in 1965, but the family did not start working it until 1976 when the household head retired. Two and one-half hectares in size, it is primarily dedicated to the production of lemons, wine grapes, and vegetables. Of these products, only the first two are commercialized, generating a net farm income of 3 million lira. The vegetable production, 5 percent of the lemon production, and 10 percent of the grapes are used entirely for personal consumption by the families of their two sons.

No financial aid is received by the farm enterprise, though a request for price support payments is pending at the EEC level. A few investments have been made in the past five years, among them the purchase of a small tractor and some equipment for irrigation. No plans for future investments have been considered.

The family's attitudes toward the farm and farm life are rather positive, as indicated by both husband and wife in their accounts of their agricultural activity. They note that the farm provides an important contribution to the family income, though this income alone is not sufficient to support the family and will most likely not be sufficient to the future either. The farm's economic relevance is not the family's principal motivation for remaining on the land. More important is the central role that it plays in the couple's life. Both husband and wife emphasize that farming is an activity that will fill their retirement years. The land itself will put them in direct contact with nature and allow them a way of life in which traditional and more simple values prevail. The farm represents, then, a new and fundamental interest for their present and future existence.

Finally, the family members declare that very few neighboring families have left the land in the last ten years, though several farms are not exploited adequately due to the fact that their owners and family members spend most of their working time attending to off-farm activities.

Third Case

The final case in this group involves a farm family of four from Ortoliuzzo county, which is located in the coastal area of Northwestern Sicily. This family is composed of a husband, wife, and two sons aged 67, 64, 35, and 32 years, respectively.

All four family members work on the farm. The husband averages 25 hours a week, while the two sons and the wife work approximately 20 hours per week. As this amount of labor is considered sufficient to satisfy the needs of the farm, no hired workers are employed.

The two sons have full-time off-farm jobs. The older is employed as a clerk by a local manufacturing company, while his brother works as a radio and television repairman. These activities provide them with economic independence and allow them to contribute substantially to the household and farm expenses. Nevertheless, they do not plan to leave the household in the near future unless forced to do so for job-related reasons. This desire to remain with their family is motivated by the satisfaction derived from living on the farm and by the fact that they do not feel that another dwelling could surpass the conveniences provided by the present one. Finally, they are not interested in forming new families of their own in the next few years.

The husband contributes to the family off-farm income as well in the form of a retirement pension received from his previous occupation. This income, together with that of the two sons, provides the family with a total income of over 40 million lira.

The farm, which was purchased in 1970, is five hectares in size and is divided into two parcels. The farmhouse is on the first of these, a three hectare plot, while the remaining two hectares are located one and one-half miles from the house. Of these five hectares, all of which are utilized for agricultural purposes, four are dedicated to the cultivation of fruit trees (peaches, apricots) and one to the cultivation of vegetables. Of these products 90 percent are sold commercially, providing the family with an annual net farm income of 7 million lira. The remaining 10 percent is consumed by the family itself. Agricultural production has been increased in the past few years by investments made for the improvement of the fruit trees and the construction of a new system of irrigation. A conversion of land presently devoted to vegetables into additional fruit orchards is planned for the future.

All of the family members are satisfied with the farm and farm life. The husband declares that he has invested in the farm so that he will have a pleasant activity in which to occupy himself after retirement. "In the beginning," he continues, "it was an occupation that I initiated without specific considerations of the economic aspects involved. I was only interested in returning to the way of life that farming provides, which was also that of my parents and of my childhood. Later, I realized that the farm could also provide a limited, yet satisfying farm income, especially considering the limited time and capital invested." He goes on to state that he views farming as the healthiest and most nature-bound of activities.

Retired Group Summary

This group of farm families is characterized by the fact that the heads of household chose farming as their principal activity after retirement. They are in many respects newcomers who prize, above all, the enjoyment of relaxation that farm life can provide. For these reasons they have elected to dedicate the final years of their life to agriculture. At the same time, the farm enterprise represents an important economic contribution to the family budget, though it would in none of the above-mentioned cases be sufficient to support the family, nor have family members ever considered it as the principal source of income. Finally, due to the retired status of all the families' senior members, active off-farm employment is limited only to the families' sons and/or daughters.

Equity Group

The fifth group is composed of farm families who remain in agriculture due to the recent purchase of a farm. It is the smallest group of all accounting for only 6 percent of the sample.

From an economic point of view, the major characteristics of these families can be summarized as follows. First, they have the second highest total family income among the six groups selected (10.5 million lira)

(Table 28). Second, their farm income is the highest among all of the groups (6.16 million lira) (Table 29). Finally, all of the farm family members in this group declare that the farm income is not sufficient to support the family, nor will it be sufficient in the future (Table 34). As far as family members' off-farm jobs are concerned, the percentage of families in which sons and daughters are employed off-farm is lower than that of families in which one of both parents work off-farm (Table 30B, C).

The farms in question are of recent acquisition, as the average year of acquisition is 1976. They also tend to be purchased rather than inherited or obtained through land reform operations (Table 32). As far as total land utilized is concerned (Table 36), their average size is the largest of all the farms in the six groups, while the percentage of family members who are satisfied with farm life is the highest among all groups (Table 26). Members of this group indicate that neighboring families left agriculture due to an insufficient farm income (Table 25). No major differences between this group and other groups were recorded with respect to farm investments.

CASES

The cases of two farm families are used to illustrate the characteristics of this group.

First Case

The first case involves a family of four living in Scaletta county in the Southeastern part of Sicily. The family is composed of a husband, wife, and two sons (65, 62, 30, and 28 years of age, respectively). Of these members only the husband and the sons work on the farm. However, the amount of time dedicated to this activity is rather low, for the sons each average 15 hours a week, while the head of the household works on the farm for only 10 hours. No hired labor is employed though all family members agree that some help is needed on a seasonal basis. The decision not to employ hired labor was made in view of the low farm budget.

All of the above-mentioned family members have an off-farm job as co-owners of a small grocery store in the nearby village of Scaletta. This activity absorbs the vast majority of their working time and provides the bulk of the family income. No other income or welfare payments are received by any family member.

The farm is six hectares in size, though only four of them are currently utilized for agricultural purposes. Purchased in 1969, it is devoted exclusively to the production of citrus fruits. This production is almost entirely commercialized, as only 3 percent of it is consumed by the family. A minor portion of the citrus fruits are sold directly to the public at the family grocery store, while the rest is sold to a middleman. However, this commercial activity has not been very profitable for the past few years. As the head of the household explains, "In the past few years the price of oranges and lemons in this area has decreased substantially. The cost of harvesting them is almost larger than the return from their sale. Now (summer, 1983) is the period to harvest lemons, but we are not going to do it because after we pay all the expenses, we will have a ridiculous profit. It is really not worth it."

The difficulties in realizing a sufficient profit have motivated the family to reduce investments in the farm. However, two major investments were made in the past few years, the construction of containers for irrigation water and the purchase of new water pumps. No investments are foreseen for the next five years.

This pessimism can also be seen in the family's attitudes toward the farm and farm life. Though family members claim to enjoy farm life, this feeling is very much diminished by the unprofitability of the farm enterprise. The situation is further aggravated by the fact that their original motivation for purchasing the farm was to invest the family's capital. All of the family members agree that the utilization of this capital for agricultural purposes has turned out to be among the least profitable. However, the family does not intend to sell the farm nor to halt the cultivation of the land. "It would mean," the head of the household states, "either selling at an unremunerative price or letting the citrus fruit trees die. In either case the value of the farm and, consequently, of the original investment would depreciate."

Second Case

The second case in this group involves a farm family living near the town of Barcellona in the Northwestern portion of Sicily. It is a family of four: husband, wife, daughter, and son, ages 59, 48, 23, and 17, respectively.

The husband is the only family member who works on the farm, averaging 15 hours a week. He is assisted in specific operations such as pruning and harvesting by hired labor. Two to four such workers are employed during the spring and fall for a total of 40 working days and are paid an average of 40,000 lira per person a day.

All family members have off-farm jobs in the family furniture shop. This activity serves as their principal occupation and source of income, providing 80 percent of the total family income.

The farm, which was purchased in 1974, is two hectares in size and is used entirely for the cultivation of citrus fruits. This production is mostly commercialized, with only 2 percent consumed by the family. It provides a net farm income of 8 million lira. There have been a few investments on the farm in the last five years, generally related to the replacement of old trees, with no plans for other major investments in the near future. There are also a few pieces of machinery present, among them a tractor and irrigation equipment.

Both the head of the household and other family members express satisfaction with the farm and farm life. although this income accounts for only a fraction of the total family income, the husband is convinced that, with the acquisition of a few more hectares of land, farm income will be sufficient to support the entire family. However, the purchase of the farm and the present farm activity are not primarily related to the family's satisfaction or appreciation of farm life. Rather, they have to do with the need to employ capital obtained from other economic activities. Given the high level of inflation (16 percent) that has characterized the Italian economy in the last decade and the relatively low rate of interest, the family was motivated to employ capital in an investment that would protect the value of the capital itself from depreciating. This strategy, according to all family members, was correct. In fact, in spite of the crisis of agricultural production, the decrease in price of agricultural commodities, the small farm size, and the limited amount of investment, the farm income is considered to be relatively high.

Equity Group Summary

This group is composed of families who have chosen to remain in farming due to an investment of capital in the purchase of land. The two cases illustrated present examples of farms that follow the above-mentioned pattern but differ substantially as far as the outcome of their investments are concerned. The first case involves a farm that does not generate any profit for the family and that is kept in business in order not to depreciate further the value of the capital originally invested. In the second case the farm, though small in size, produces a relatively high profit which signifies a valorization of capital.

The different outcomes of the investments do not seem to generate differing attitudes toward the families' willingness to remain in farming. In the first case emotional reasons further motivated the family to stay on the land. One member of this family states: "Many farms have been abandoned and once beautiful plantations have been destroyed. We do not intend to follow the same pattern. It would be a shame." In the second case the partial success of the investment and the small role that the farm plays in the overall economic activity of the family strengthen the family's intention to remain in farming.

Residual Group

The sixth and final group, the Residual group, consists of several subgroups of families that remain in agriculture for differing reasons. Among them three assume particular theoretical relevance. The first consists of families that are ready to sell the farm and will do so at the first available and convenient opportunity. The second subgroup is composed of families who would like to sell the farm but consider the offers received not adequate for the real value of the farm. The third is the subgroup of families in which farming is considered a hobby and/or a recreational activity.

The decision to combine into one group farm families who express such differing motivations for remaining in agriculture is based on the limited size of each of these subgroups. While there is a significant difference among them from a qualitative point of view, their quantitative size is too limited to generate statistical values in which one can place much confidence. Consequently, the analysis of this group will be limited to qualitative aspects and no statistical comparisons with the other groups will be undertaken.

CASES

Three cases have been chosen to illustrate the characteristics of the major subgroups of the Residual group.

First Case

The first case involves a family of two, a husband and wife, 66 and 61 years of age, respectively, who live in St. Margherita county north of the Sicilian town of Milazzo.

Both family members work on the farm with an average of 40 hours a week for the husband and 20 hours for the wife. This amount of labor is considered sufficient to perform all farm tasks, so no hired workers or additional help is required. Though farming is currently the only working activity of the couple, they had off-farm occupations from which they retired a few years ago. From these off-farm jobs they now receive pensions that constitute 95 percent of the total family income.

The farm is two hectares in size and is divided into two parcels, which are located one mile apart. The first of these is one-half hectare in size and contains the farmhouse, while the second contains the remainder of the land. Since inheriting the land in 1960, the couple has been engaged in agricultural activities first as part-timers and then on a full-time basis after retirement. The principal farm production is that of lemons, which are sold almost entirely to a middleman, with only 5 percent consumed by the enlarged family. However, due to the low market price for this crop and the particular conditions of the selling agreement with the middleman, the net farm income derived from this commercial activity does not exceed 1 million lira. The low returns and uncertainties of the market have motivated the family to limit investments to a minimum in the past five years and not to plan any in the immediate future.

The family's attitudes toward the farm and farm life are rather negative. Both family members declare themselves to be dissatisfied with the farm, their major source of disenchantment being their inability to realize an adequate income. They stress that the amount received from this activity is far too low and certainly not remunerative of the time and capital invested in it. These considerations have motivated many neighboring families to sell their land, a solution that the family would like to adopt in the immediate future. Moreover, conditions are becoming more favorable for the sale of the land, as its price is constantly escalating. In the past few years the area has been the site of numerous projects of land development aimed at the construction of residential dwellings serving both the area of Messina and Milazzo. Consequently, the family intends to sell the land as soon as a profitable offer is made to them. This might possibly occur within the next one or two years.

Second Case

The second case involves a family of four living in Altolia county in the Southwestern portion of Sicily. The family is composed of a husband, wife, son, and daughter (35, 34, 10 and 7 years of age respectively). The husband is the only family member who works, average 30 hours per week on the farm while simultaneously working as a clerk at the National Telephone Company. This off-farm job provides the bulk of the total family income (95 percent).

The farm was inherited from an uncle in 1973. Since that time the two hectares that compose it have been employed for the cultivation of citrus fruits and vegetables. The citrus fruits (90 percent lemons, 10 percent oranges) are almost entirely commercialized, while all of the vegetables and the remaining portion of the citrus crop (5 percent) are consumed by the family. The commercialization of the citrus fruits is accomplished through the mediation of a cooperative, which purchases the farm production of all its members but pays for only a part. The remainder is paid after the product has been sold to the public through the cooperative store or other stores. Through the years the cooperative has been effective in boosting the prices of commodities sold but has not been able to avoid the general decline that these products have experienced in recent years. Due to low market prices and the limited production generated by the farm, its net income does not exceed 2 million lira.

The low farm income is largely responsible for the dissatisfaction that family members express with respect to the farm and farm life. On more than one occasion the husband has considered selling the farm but feels that no appropriate offers have been made. He states: "We are willing to remain on the farm until a good selling price is found. The proposals that we have received are not adequate for the real value of the land, and we feel it is better for us to remain." Moreover, family members are opposed to the solution of abandoning the farm that has been adopted by many neighboring families. In their estimation, such an option would be inadequate, as the economic and non-economic advantages provided by the farm would be lost without receiving tangible compensation.

Third Case

The final case of this group involves a household located in Mammola county in the Western portion of Calabria. It is composed of a husband, wife, and two sons. The ages are 47, 45, 14, and 12 years respectively.

Unlike the other cases examined thus far, the bulk of work done on this farm is not performed by family members. Though the husband averages 10 hours a week, a farmworker employed year round and other workers hired for shorter periods carry out the vast majority of the farm duties. Both husband and wife hall full-time off-farm jobs, which provide the family with 90 percent of the total annual income. He is a bank executive, while she teaches in the local high school.

The farm is four hectares in size. In 1979 it was inherited by the family and has been employed for the cultivation of oranges (1 hectare), vegetables (1 1/2 hectares) and olive trees (1 1/2 hectares) since that time. In spite of the relatively large portion of land devoted to vegetables, none of these products is commercialized, rather they are consumed by the family. The orange crop, 5 percent, and 10 percent of the olive production, which is converted into oil, are also retained by the family. The remainder of the products are sold to local grocery stores, providing a net farm income of 3 million lira, which is also complemented by price support payments received from the EEC. In past years there have been a few investments on the farm. The most important of these has been the substitution of some citrus fruit trees and the replacement of old irrigation pumps. The construction of a new storage room for olives is forecasted for the future.

The family's attitudes toward the farm and farm life are positive, with the husband manifesting the highest degree of satisfaction among all family members. He considers farming the most effective way to be in contact with nature and to avoid the routine of a town job. He adds; "In my case farming is not a real economic activity but a hobby, which I enjoy very much and would like to continue in the future." The other family members are also satisfied with living on the farm in spite of their present and future lack of involvement in farm activities.

Residual Group Summary

The Residual group is composed of several subgroups of families that remain on the farms for differing reasons. Among them three have been considered for their theoretical relevance. In the first description the farm is in the process of being sold, and the family is ready to leave the sector in the immediate future. In the second case, in spite of the family's positive attitudes toward selling, the farm has not been sold due to the fact that no remunerative offers have been made. In the third and final case farming is considered a healthy and desirable hobby by family members. In all of the above-mentioned cases, the farm enterprise plays a minor role in the overall economic situation of the family. In the first two cases the sale of the farm is the most important concern, and it is the principal condition for their temporary presence in the sector. In the third case all economic values associated with the farm are considered secondary in relation to cultural and emotional aspects.

CONCLUSION

The six groups include farm families with differing motivations for remaining in agriculture. There is, thus, no single common denominator to account for their persistence in the sector but, rather, a complex pattern of varied and contradictory elements. Though this pattern cannot be explained by any single theory[14] developed by classic and modern sociologists, it can be interpreted by a combination of such theories. This pluralistic posture thus constitutes the interpretative framework for the analysis of the groups and cases.

A number of cases treated in this study may be interpreted in light of Weber's account. For Weber (1958) the persistence in agriculture of small holdings was related to the farm family's search for independence based upon a voluntaristic act. His description of this phenomenon refered in particular to the situation of East German farm families and farmworkers toward the end of the last century. In this context Weber underlined the differences existing between peasants, who are still bound to the declining feudal productive system, and small farmers, who are owners of their land and their enterprise. He emphasized the willingness of the latter to endure extreme sacrifices in order to obtain and then preserve their land. Giddens summarized the phenomenon: "The worker who possesses his own small plot of land will endure the most extreme privations, and the most heavy indebtedness to usurers, in order to preserve his 'independence'" (1971:123). According to Weber, such an action is based, first of all, on the desire to escape the domination of the patriarchal system of property and production (feudalism). Second, it represents an attempt to contribute to the establishment of a system of free enterprise and free competition as the principal mode of production. This search for freedom and economic independence is ultimately motivated by a set of new values that developed outside of and in opposition to the preexisting feudal order. These values convey a novel social vision that finds expression in the strong attachment to the land and in the new concept of rural life displayed by farmers.

In one sense a parallel can be drawn between the Weberian assumption and the Jeffersonian ideal, which was developed a century earlier and constitutes one of the major motifs of American society. Both stressed the need for free enterprise and free will of the actor over the economic and social restrictions of a system based on serfdom. Nevertheless, the defense of these (capitalist) values becomes a dominant theme in the Jeffersonian ideal, while such a position cannot be found in its entirety in the Weberian account, for Weber emphasized that the possibility of such a "freedom" is merely an illusion (Giddens, 1971:123). However, Weber continued, this illusion must be taken into serious consideration if the human activity is to be understood. The values attached to this "freedom" ultimately represent the motivations that explain behavior that could not otherwise be explained by an economic or materialistic understanding of social relations. The fact that the independent farmer chooses to be remunerated less than a farmworker or to loose all the communal advantages granted to a serf-peasant by the feudal system is a phenomenon that is not easily explicable from the economic-materialistic point of view. Yet, in light of the values mentioned above, it is not only understandable, but it can also be seen as a means for social change and progress.

The remarks made by Weber almost a century ago are largely applicable to the motivations of some small Southern

Italian farmers for remaining in agriculture. Members of the traditional group express the same orientation toward the farm, the land, and working under the supervision of others, (see also the first case in the Reserve group) that Weber describes for German farmers. These family members' under-standing of "love toward the land" (see the second case in particular) and their drive for independence in work relationships (third case) are all cases in point. However, this writer must depart from Weber's account on two fundamental issues. First, the epistemological posture used by Weber in analyzing the relevance of values in the explanation of social action is extraneous to the dialectical approach to sociology. For Weber, though he denied that social action can be interpreted as a simple derivation of values and economic factors alone[15], the relationship between structure and superstructure is ultimately that of direct or indirect causation. In the dialectical account any relation of causation is denied, as the material relations of production (thesis) and ideology (antithesis) are two sides of a single process whose outcome (synthesis) cannot be know <u>a priori,</u> but derives from the interrelation of thesis and antithesis. Second, as a dialectical reading of both the German and Italian situations suggests, it is important to note the different meanings that such values have in the two historical periods in question. In nineteenth century Germany, farmers' attachment to the land was (as also recognized by Weber) a progressive position favoring the rejection of dominant values supportive of a pre-capitalist system based on serfdom. Today in South Italy (and in the United States as well), such values are the expression of a rather conservative position, for they support an established and dominant ideology. In other words, the farmers' allegiance to these values is legitimative of the social and political system as a whole.

Such considerations become extremely useful in explaining the relationship between farm income and feelings of love and attachment toward the land expressed by some family members in the sample. In a number of instances (Traditional group and in some cases of the Reserve group), the attachment to the land and to agrarian values can be interpreted as a result of a family member's inability to acquire an urban standard of living, which is understood to be economically more remunerative and socially more gratifying than a rural one. As the former cannot be achieved (due to a lack of job alternatives that prevents upward mobility, and/or low farm income that prevents social upgrading, etc.), agrarian values come to legitimate a status quo in which farm families are in a marginal social and economic position. A similar sharing of agrarian values is also present among families in the retired group. Although the decision to choose farming as a post-retirement activity has been described in some analyses (Berry, 1979; De Janvry, 1980) as a rejection of the dominant urban lifestyle and values, a different understanding appears to be prominent in the cases illustrated above. For these families retiring on the land is understood, above all, as a return to a traditional lifestyle and to values which are considered more desirable than others.

The Weberian understanding of the persistence of small farmers in agriculture has been employed and developed by Brusco (1979). As mentioned above, Brusco saw the persistence of small farms as generated by voluntaristic acts based on a value-oriented understanding of the social situation by the farmer. He also explained the phenomenon as the consequence of economically motivated attitudes aimed at increasing the total family income. While the limitations of such an interpretation have been identified above (Chapter 1), it is important to stress the validity of this theory in explaining the motivations of some farm family members for remaining in agriculture. The Complementary group offers a good case in point. In that group farm families do not leave the sector because their members view the farm income and other economic benefits generated by the farm enterprise as a complement to off-farm earnings. As in the same pattern illustrated by Brusco for Northern Italian farmers, they are willing to continue farm work in order to produce extra income, regardless of the level of productivity of labor and land or the hardship of agricultural life.

The Kautskyan account of the persistence of small farms and the modern explanations inspired by it are also extremely relevant to an understanding of the Southern Italian case. Kautsky stressed that the persistence of small farms is linked to the role of producers of surplus labor that such farms perform within the capitalist mode of production. This labor force, which is not fully utilized on the family farm, is then employed at low cost by capitalist enterprises that can decrease their production expenses and accelerate their process of accumulation. While such an explanation cannot be employed to account for all of the cases studied, it finds immediate support in a large portion of farm families that have at least one member engaged in off-farm activities. As with the farmers in

Kautsky's analysis, these family members are largely forced to find an off-farm job. The insufficient farm income and the gap between supply and demand for labor on the farm makes, as Mottura and Pugliese (1980) suggested, a portion of the farm family members "unemployed on their own farms" and forces them to seek job alternatives. Moreover, the accuracy of this explanation is supported by the fact that over 75 percent of all farm families have members involved in off-farm activities. Such families are distributed across the six groups considered in this study.

Calza-Bini's elaboration of Kautsky's thesis provides a further explanation with regard to farm families classified in the Reserve group. Calza-Bini (1974; 1976) indicated that the persistence of small farms is related to a lack of jobs outside agriculture. Consequently, the farm becomes a place where family labor remains partially unemployed and/or underemployed while waiting for job alternatives. Numerous farm family members of the Reserve group remain in agriculture for precisely the reasons stressed by this author. Their inability to obtain remunerative jobs outside the sector, their lack of professional skills, and the precarious situation of the labor market, force them to choose farming as their principal source of income, though it is largely insufficient to provide an adequate standard of living.

However, Calza-Bini emphasized another element that can be applied to families classified in several of the groups studied (Traditional, Complementary and some families in the Reserve group). He indicated that while there is a lack of stable full-time jobs, the decentralization of the industrial productive system in Italy has created a number of part-time, low-paying jobs that are available to farm family members. Accordingly, a relationship of mutual support is created between the farm sector and decentralized industrial activities. Industrial decentralization is possible due to the availability of surplus labor kept on the farm, while the farm enterprise can continue to exist as a result of the part-time jobs available to family members. Another aspect of this relationship is that neither the industrial nor the agricultural labor demand can provide full employment to a substantial portion of the available workforce. Consequently, a situation is generated in which both jobs are needed; the incomes complement each other, and a low standard of living is maintained. Several of the cases indicated in the first part of this chapter fit this explanation.

Finally, it must be added that Calza-Bini's account seems to contradict Brusco's explanation of farm families' motivation for integrating off-farm and farm income (Complementary group). This contradiction is, however, more apparent that real, for both theories refer to phenomena that occur simultaneously and are two aspects of the same process.

Daneo (1969) also makes a contribution toward the explanation of the persistence of small marginal farms in South Italy. He emphasized the role that the State plays vis-a-vis the payment of welfare benefits to farm family members. As observed in cases in all of the six groups identified, State-sponsored welfare benefits are received by a substantial proportion of the population. This share represents 62.3 percent of all families in our sample. Furthermore, another 42.5 percent of farms in the sample receive financial aid from national or international organizations. Although it is difficult to say that State aid, whether direct to family members or to the farm enterprise, can account alone for the persistence of the marginal farm, it is evident that such economic support plays an important role in their existence. Daneo's account of an extensive welfare system that pays benefits to a seemingly large number of recipients is thus of considerable relevance to this study. For Daneo this action is taken by the State in order to maintain a political and social equilibrium in society (legitimation). It also generates the persistence in agriculture of families whose farms could not survive depending only upon their own economic resources.

These observations emphasize an important and closely related aspect of the persistence of small farms, namely, their connections to the overall social system of which they are a part and to the mechanisms and forces that contribute to its reproduction. However, the cases included in the sample deny Daneo's rather functionalist final conclusions. In these cases the contradictory aspect of the process of State intervention in agriculture and the relationship between the State and poor farmers appear clear.

As illustrated in the first case of the Traditional group, the relationship between farmers and the State in South Italy is one of "dependence and conflict" (Pugliese, 1983). The farmer needs State support, a need that is particularly strong in two instances. In the first of these, the family runs a marginal farm that produces a very low or nonexistent income and no other sources of income are available. The assistance of the State is thus needed to relieve the family's high degree of indigence.

In the second case the farm is productive and operates competitively within the market. However, in order to provide a sufficient income for the family, the farm enterprise must receive financial aid. When effectiveness of State support is reduced by a worsening of general economic conditions or by the reduction of state expenditures in real terms caused by a fiscal crisis (O'Connor, 1973), the reaction of the farmer is that of identifying the State as an antagonist from which a solution to the problem is demanded and the blame for the malaise attributed. As Pugliese (1983) emphasized, this reaction is related to the historical role that the State has taken in South Italy over the centuries. His conclusions, which are largely confirmed by statements of farmers interviewed in this research, stress that the State is seen as an institution that has traditionally taken resources away from farmers, and, consequently, its primary duty today is that of repaying them in the form of welfare benefits. This creates a situation of perpetual dependence on State aid and, simultaneously, of antagonism in which farmers and the State are opposing factions. In the case of the Traditional group mentioned above, this situation is summarized in the words of a head of household who states that the only alternative he and his family have to survive in farming is that of an increase in State support.

Consequently, the conclusions reached in the present study point toward a situation of legitimation of the political and social status quo on the part of small marginal farm families. At the same time this process is countered by the conflict that such farmers experience with the State. The State is thus forced, in order to reduce the level of this conflict, to maintain continually and even to increase its economic intervention in the farm sector. This, in turn, may be a difficult task to perform if a shortage of funds is experienced by the State.[16]

The relevance of small marginal farms in the overall socio-political system also has been emphasized by Mottura and Pugliese (1975, 1980), and their conclusions have been confirmed by several cases in this study. As mentioned in the first chapter, Mottura and Pugliese stressed the dual role played in advanced societies by the small farm sector. The first of these is the productive one[17], and the other is that of keeper of surplus labor. This second role, according to the two authors, is dominant among small farms in periods of recession or in situations in which accumulation is not possible. Farm families who remain in agriculture due to a lack of job alternatives and farms that provide an economic complement to the families' other incomes thus support the conclusions of Mottura and Pugliese.

It should be noted that the productive role of small farms in our sample is extremely limited, as only a few can support the family without income from outside sources. The explanation of this phenomenon lies in the organizations of the agricultural sector and market, which do not allow an adequate monetary realization of the already low production (Castellucci et al., 1984). It can be found also in the overall poor physical characteristics of the land (mostly mountainous and hilly) and in the farm's limited physical size. Furthermore, the lack of economically self-sufficient farms tends to deny the application of Levy's (1911) hypothesis of the persistence of small farms to the case of South Italy and Sicily. Although almost all of the farms that do provide a sufficient income to the family are devoted to intensive cultivation (which is the principal reason for their survival, according to this author), there is no evidence that they can out-compete larger farms. In fact, as several other studies have also pointed out (Castellucci et al., 1984; Fabiani, 1979; Mingione, 1981), various important factors lessen their chances for competing with larger farms. First, the distorted system of commercialization of agricultural products in Italy needs to be called into question. This system is largely controlled by middlemen who receive a share of up to 60 percent of the final price of agricultural commodities, which leaves the producers with a share of less than a third of the retail price. Second, the costs of up-to-date technical and mechanical inputs are generally too high for small farmers, who are forced to forego most of them, thus further increasing the productive gap between large and small farms. Finally, large farms receive the majority of available domestic and international financial aid, which allows them to reduce further costs of production and prices without altering profits. Such an option is almost unavailable to small farms (see also Chapter 3).

It should also be noted that while small farms cannot outcompete large farms, in some instances their presence influences the economic behavior of large producers. More specifically, the fact that small farms in the region under study produce at competitive low prices prevents larger producers from selling at inflated prices, thus forcing them to follow the trends of the competitive market. This phenomenon is limited, however, only to local markets (markets close to the area of production available to small farmers) and does not involve larger national and international markets, which are not accessible to small producers.

The abovementioned theories pertain to cases found in all of the six groups identified. However, the majority of the cases in the equity and residual groups are characterized by patterns not clearly identifiable by such theories. Further treatment of these groups is thus needed.

The principal characteristic of farm families belonging to the fifth group (Equity) is the fact that they have invested in the land and refuse to leave the farm in order to protect this investment. In all of the cases mentioned above, the purchase of the land was considered a very sound investment and a good employment of available family capital. As Mingione (1981) has indicated, this attitude has long been in existence among Southern Italian farmers and is based on the success that land investments have enjoyed since the beginning of the century and particularly from the end of World War II to the sixties. During this period a large portion of agricultural land was subject to urban development, which generated a substantial increase in its value and made many farmers rich almost overnight. The deceleration in the process of urbanization and the crisis of the construction sector that Italy has experienced almost continuously since the late sixties have limited sharply the profitability of such investments, which frequently turn out to be clear losses. A second factor, which is related to the unstable conditions of the italian economic system (De Benedictis, 1980), also has been identified for the popularity of land investment. Periods of high inflation, from the rampant inflation of the post World War II era to the double-digit inflation of the seventies, made land investments especially attractive among individuals with a rural background or with interests in farming.

In the cases of farm families illustrated, the popularity of these beliefs with respect to land investment appears clear despite evidence against their profitability. However, the ultimate motives for the persistence of these families in agriculture are related more to material economic reasons (the desire to increase or maintain the value of invested capital) rather than to unchallenged beliefs. The latter influenced the entrance of these families into the agricultural sector but are not primary motivations for their persistence.

The final group, Residual, is composed of farm families who remain in agriculture for differing reasons. Some of them, as illustrated by the first two cases, are families in the process of abandoning the sector whose reasons for remaining are related to the conditions of the land market. In the first case these conditions call for immediate sale of the land due to urban development in the area, while in the second case the exclusively agricultural nature of the land has not generated a price high enough to satisfy the family's demands. It is important to mention at this point the distorted condition of the land market in South Italy. Despite the desire of several landowners to sell their property, the expectation of high profits generated by the beliefs presented above creates a gap between supply and demand. This gap is widened by the overall economically depressed character of the region, which does not produce a consistent and sufficient demand. Consequently, the market is rather stagnant, especially for land that will not be developed in the near future, and prices are kept high at levels twice those found in the United States (De Benedictis, 1980; Diaz et al., 1983). As a result the exit from the sector of the family in the second case may be delayed for a longer period than expected or desired by family members.

The third case is rather different, for the farm family's persistence in agriculture is motivated by feelings of attachment to the land. These feelings differ from those of families in the first (Traditional) group, however, as they express a position that is critical toward dominant urban values and patterns of social action. Their statements indicate a perception of farm life as a break from the everyday routine and as a healthier manner of spending leisure time. Consequently, rather than sharing the dominant societal pattern and legitimating it, these families support an alternative view.

The different family and farm traits that have been identified and discussed in the preceding pages do not deny the existence of similarities among all of the six groups. First, a very relevant common denominator of the entire sample is the fact that farm families have at least one off-farm source of income. Out of the total 199 families sampled, only 24 percent of them have neither a family member working off the farm nor a member retired from off-farm activities, but they all receive welfare payments under various programs.[18] Second, although farm income contributed to the economic well-being of most of the families and is considered indispensable by a significant portion of them, the majority of the farm families in the sample receive the larger portion of their income from off-farm activities and cannot maintain an adequate standard of living without such economic input.[19] Consequently, it can be argued that the persistence in the agricultural sector of these farms is not primarily dependent on farm income. In other words, and paradoxically, these farms remain in agriculture because the economic well-being of the family is _not_ primarily related to farming as a gainful enterprise. Finally, as will be explained in greater detail below, a substantial portion of small farms do not accumulate. They do not produce a sufficient monetary return to guarantee the survival and reproduction of the labor force living on the farm. This absence of accumulation raises the question of their position within the social structure. In Chapter 2 and Chapter 5, the action of social legitimation performed by small farms and the small farm sector within society was indicated. The connection among lack of accumulation, legitimation, and persistence of small marginal farms is in need of further analysis. This is the topic of Chapter 6.

6
Accumulation, Legitimation, Small Farms and the State

ACCUMULATION AND LEGITIMATION

Accumulation, according to the Marxian tradition, refers to the process of reproduction of capital. More specifically, it involves the reconversion of surplus-value into capital; that is, the employment of a portion of the revenue as capital (Marx, 1972:365)[20]. In order to accumulate, though, it is necessary to obtain sufficient revenues (surplus-value) to cover the costs of production and then employ the remaining surplus-value for the purchase of additional means of production (labor, raw materials, and machinery). As Marx explained:

> To accumulate, it is necessary to convert a portion of the surplus-product into capital. But we cannot, except by a miracle, convert into capital anything but such articles as can be employed in the labor process (i.e., means of production), and such further articles as are suitable for the sustenance of the laborer (i.e., means of sustenance). Consequently, a part of the annual surplus-labor must have been applied to the production of additional means of production and subsistence, over and above the quantity of these things required to replace the capital advanced (1972:366)."

It follows that accumulation is guaranteed by three related processes. First, it is necessary to provide a means of subsistence to labor; namely, wages sufficient to guarantee the survival and reproduction of labor must be generated through the economic cycle. Second, the costs for the purchase and use of technical elements and raw material necessary for production must be covered. Finally, a profit must be generated and invested in the economic cycle. In the event that one of these processes does not take place, accumulation cannot continue to occur.

As illustrated in detail in the exposition of the characteristics of the farms considered in this study, the processes necessary for accumulation are for the most part absent with respect to these farms. Consequently, it can be argued that the vast majority of them do not accumulate. As indicated above (Chapter V), a large portion of the farms (83 percent with representatives from all of the groups identified) cannot provide an income adequate for the subsistence of the family. Moreover, a large portion of them (64 percent) provide little or no income to the family. These characteristics constitute a violation of the first requirement for the existence of accumulation; that is, the inability to generate wealth to guarantee the subsistence of labor.

Second, the lack of farm income signifies a lack of farm profit; namely, farm revenues are not sufficient to cover other costs of production besides labor. Finally, a lack of profit makes it impossible for a considerable portion of these farms (58 percent) to invest, which constitutes a violation of the major condition for the existence of accumulation: the reutilization of surplus-value in the form of capital.

There are other characteristics that apply to a number of small farms in South Italy and Sicily, which can be used as evidence to support the thesis that there is a low level of accumulation among these agricultural units. More specifically, the fact that most of the farmers considered are part-timers while their off-farm job is the principal form of employment indicates an ongoing process of proletarianization of this labor force. Farm family members tend to lose their position as agricultural petty bourgeoisie, that is owners of both their means of production and labor[21]. Consequently, through proletarianization the petty bourgeoisie gradually relinquishes its ability to control and fully use the means of production, thereby losing its potential for accumulation.

Another characteristic that indicates a non-existent or low level of accumulation among small farms is their contribution to total national agricultural production. In almost all of the advanced Western societies, the bulk of agricultural production is generated by large farms (Fabiani, 1979; GAO, 1978). Small farms, in spite of their large number (in Italy they account for 85 percent of all farms), produce a relatively small percentage of total agricultural production. In Italy the contribution of large farms in the last twenty years has increased to 60 percent, while that of small farms, though very high during the fifties (Fanfani, 1977), today accounts for only 30 percent of the national figure. This means that 30 percent of the total agricultural revenue in Italy is divided among 85 percent of the total of the farms, while 60 percent of all agricultural revenues is controlled by only 12 percent of the farms (the remainder is controlled by farms of medium size). In other words, the reproduction of wealth (accumulation) is much higher in the large farm sector than in the small farm sector.

It can be argued at this point that accumulation is the most important element in the process of reproduction of capitalism. However, it is by no means the only element needed for the existence and reproduction of the system. The need for the existence of other elements can be explained as follows.

Under capitalism the primary interest of the bourgeoisie is generating profit. Profit generation, in turn, becomes the main characteristic of the system itself. Everything that forms a part of the profit-generating system should therefore be represented under the commodity form. As long as every value (goods and labor) can successfully be turned into a commodity and exchanged in the market, capitalism will develop prosperously (Offe and Ronge, 1979). Accumulation thus takes place as long as every value appears in the form of a commodity (Offe and Ronge, 1979:348).

However, it is not always possible to transform every value into the commodity form. History has shown that in various periods of capitalist development, crises arose that paralyzed the commodity form of value. Under capitalism there is no certainty that each value offered for exchange is actually exchanged, nor is there any assurance that each value is transformed into a commodity. This process, nevertheless, should be self-corrective (Say, 1834; Smith, 1976). That is to say, when a value is not exchanged, the owner of the value is forced to reduce the price to the point at which a demand is generated (the value is exchanged). When this occurs, the value is once again turned into a commodity and the process of accumulation can take place. In periods of crisis the failure of values to be transformed into the commodity form is overcome by the reduction of their prices and the subsequent generation of demand. Say (1834:82) argues that: "Each supply creates its own demand." The functioning of this self-corrective mechanism does not seem to be common, however. Its absence is particularly apparent in advanced capitalist societies (Offe and Ronge, 1979:349).

A large body of literature suggests this self-corrective theory to be inaccurate[22] (Keynes, 1971; Sweezy, 1942). As Offe and Ronge suggest, "There is plenty of everyday evidence to the effect that both labor and capital are thrown out of the commodity form, and that there is little basis for any confidence that they will be reintegrated into exchange relationships automatically" (1979:349). Accumulation is thus not an automatic process, but is determined by the ability of the bourgeoisie to eliminate the paralysis of the commodity form of value and push down prices when necessary[23]. It is essential, then, for the bourgeoisie to cope with any social opposition in order to reach the above-mentioned goals. Consequently, the process of accumulation requires a degree of domination of one social class over the other (Marx, 1973).

Domination of one class over another was achieved through the use of open coercion under past modes of production. A class could fully control the activity of another through the use of force under the slave and feudal modes of production. The worker was then integrated into the means of production. The slave was an instrument of labor, while the serf was "integrated" into the land (Perez-Saiz, 1981).

> This implies that the class owning the means of production has to consider the laborer as part of its material condition of existence. In this sense, to make the property effective, this class has to appropriate an alien will: the laborer's. This appropriation is only possible through the exercise of open coercion. Therefore, domination in an explicit way is presupposed in the conditions of existence of slave and feudal modes of production (Perez-Saiz, 1981:131, emphasis added).

Under capitalism, subordination takes a different form. One of the most important characteristics of this system is the separation of the worker from the means of production (Marx, 1965). This separation implies that the worker is forced to use his/her ability to work and exchange it as a commodity for a wage in return (Marx, 1972). Because this is the case, there is no longer a process of overt domination. In fact, the relationship between the class that owns the means of production (bourgeoisie) and the class that owns labor (proletariat) is based on mutual agreements that presuppose the free circulation of labor (the right to quit a job and the right to fire). Free circulation of labor and the rights of individuals against physical coercion (slavery and serfdom) are guaranteed through legislation. As Habermas (1975) has pointed out, the abstract values of equality and freedom among men are not only elements upon which the bourgeois revolution was based but also those upon which further dominance of the capitalist class and State is extended to society. These elements, then, cannot be denied in principle in any capitalist society[24]. However, in order to have accumulation, some form of domination is needed that does not take the form of open coercion.

The mutual and free exchange of commodities in the labor market is free only in appearance. In order to survive, the worker has no other alternative than to sell his/her labor for a wage to those who control the means of production. It is evident that the action of the bourgeoisie not only creates the demand for labor but also creates its supply. The "free" exchange of commodities is in effect controlled by one class making it only free in appearance and thereby serving as a form of hidden coercion (Marx, 1973).

The separation of the worker from his/her labor force transforms him/her into a commodity alienated from the human being. The worker tries to sell the commodity labor at the highest price (wage), while the capitalist tries to buy it at the lowest price. Moreover, the capitalist seeks to use the labor he/she has purchased as effectively as possible, a motive that is resisted by the worker who tries to preserve the commodity labor as much as possible in order to re-sell it for a higher price and/or for a longer time. This situation creates an opposition between two guaranteed rights: (1) the right of the capitalist to buy and use the labor as efficiently as he can, and (2) the right of the worker to preserve his/her labor power.

However, accumulation requires a constant effect on the part of the bourgeoisie to reduce the opposition of the working class, in this case an attempt to control the proletariat's right to the use of its labor. As Marx has said, "Between equal rights force decides" (Marx, 1972:344). This force, however, cannot be open but must be a covert coercion that appears legitimate, given the basic principles of society. This process of constructing a justification for an act of coercion in capitalism is referred to as legitimation. Put in a broader context, legitimation is the attempt to eliminate tension in the system in such a way that accumulation is possible. Legitimation, therefore, represents the reproduction of the conditions of social harmony that allow for the accumulation of capital (O'Connor, 1973; Offe, 1973; Poulantzas, 1978).

Accumulation and legitimation must be seen in relation to one another and not as separate processes. Their dialectical interrelationship is indispensable for the existence of capitalist societies. These processes operate at two levels, the societal and the sectorial. While accumulation has always been analyzed at both levels, analyses of legitimation tend to link it exclusively to the role of the State, which is usually viewed as the only source of the hidden coercion needed for accumulation. The case has been argued in the following manner:

> The liberal State (was) engaged in a continual process of upholding the principles of freedom and equality, while constantly modifying their application in practice in order to overcome the contradictions continually created by ... the relations of production (Holloway and Picciotto, 1977:89).

Following this perspective, analyses of legitimation were for the most part oriented toward the societal level, thereby assuming a rather holistic posture. Later analyses emphasize the delegation of legitimation that occurs in capitalist societies, a process related to the diversity of roles assumed by the State through the action of its officialdom.

As indicated above, in advanced capitalism the process of accumulation is delayed by a series of intervening factors that call for increased State involvement in the economy (at the theoretical level this process is embodied in all the Keynesian economic theories of development). The State, as Offe (1973) and O'Connor (1973) clearly pointed out, is forced to try to remove these obstacles and accelerate accumulation. In order to achieve this goal, the State has two options for intervention (Offe called them "allocative" and "productive", while O'Connor spoke of "social capital expenditures" and "social expenses of production"), which in some instances call for the involvement of other social "spheres."

The first option of the State is that of using its authority to accelerate accumulation, i.e., using resources and powers that intrinsically belong to it in the economy. These resources and powers are the right to collect taxes and employ them and the right to select the manner in which such revenues will be employed. The State authority to "allocate" and "employ" these resources is automatically legitimated by the political power invested in the State itself, that is, by the power to make ultimate decisions that belong to the State.

The second State option involves action that goes beyond the State sphere and penetrates other sectors of society. Legitimation is, therefore, also delegated to other "spheres" of society. In this case the State provides some physical inputs into production such as the recreation of labor skills via vocational programs (Keane, 1978), the construction of infrastructure such as highways (O'Connor, 1973), and the distribution of economic and productive incentives to private enterprises (monetary grants to stimulate production, discounts for loans, price support programs, etc.). In addition, the State itself becomes an entrepreneur. It directly invests in production, both in joint ventures with private capital (the enterprises created by this State activity are known in social democratic countries in Europe as "Enterprises with State Participation") and/or as sole investor.

In all of the forms in which this "second option" is carried out, the process of legitimation is not limited to the State sphere itself but is extended to other spheres of society linked to the State by State intervention. Consequently, not only State institutions but also religious, military, and economic ones serve to legitimate the capitalist system (Carnoy, 1984; Jessop, 1983; Offe, 1973; Poulantzas, 1978). The case of price support programs in agriculture provides an example in which legitimation is displaced from the State (as sole legitimative actor) to the farm sector.

CONTRADICTIONS IN THE PROCESS OF ACCUMULATION AND LEGITIMATION

The relationship between accumulation and legitimation has been described thus far as indispensable for capitalism. Its existence, however, is not a given in the system but is dependent upon the level of class struggle (Castells, 1980; Mottura and Pugliese, 1975; Wolfe, 1977). This characteristic makes the relationship between accumulation and legitimation dialectical and contradictory.

The contradictory aspect of this relationship can be described as follows: first, it is important to specify the distinction between the ruling class and the State. In some early Marxian analyses that identify with the instrumentalist school (Carnoy, 1984:104), the State is described primarily as an "executive committee of the bourgeoisie." This interpretation, which is rooted in some ambiguities in Marx's own writings on the State (Offe and Ronge, 1979; Piccone, 1983), implies a total identification between the two entities so that the State is nothing more than an extension of bourgeois power. Later analyses (Block, 1980; Habermas, 1975; Offe, 1973; Poulantzas, 1978), however, reject this position, claiming the continuous existence of tensions and conflicts between both the bourgeoisie as a whole and State officialdom (bureaucracy) and between fractions of the bourgeoisie and the State. These analyses also point out that the ruling class cannot be considered at any time a uniform entity.

This second position, which is today widely accepted, implies that while the objectives of the ruling class and those of the officialdom of the State are similar, they are also fundamentally different. The primary objective of the ruling class is the pursuit of profit, while the main objective of the State is the protection and sanction of "a set of rules and social relationships which are presupposed by the class rule of the capitalist class" (Offe and Ronge, 1979:346). In other words, the State does not protect the interests of the ruling class, but guarantees the survival of the entire system.

The similarity between the primary goal of the bourgeoisie and that of the State is that they both tend to support the process of accumulation. They differ, however, in that while each capitalist is interested in the reproduction of his/her own private capital, the State is interested in the reproduction of capital in general.

The separation of the State from the bourgeoisie does not mean, however, that they can operate independently of one another, for the existence of the State is based upon the appropriation of a portion of the wealth created by privately owned capital and corporations. In fact, once accumulation takes place, the State can collect funds in the form of taxes that provide the finances necessary for its survival and reproduction. Consequently, the State's existence is based upon a process of creation of wealth over which it has little control (Offe, 1972).

The process of appropriation of wealth by the State is, however, resisted by the bourgeoisie. This resistance is explained by the three related phenomena. First, the funds collected by the State are directly subtracted from the process of accumulation controlled by the bourgeoisie. That is, a portion of the bourgeoisie's capital cannot be immediate reutilized by this class in the process of accumulation because it is utilized by the State.

Second, the State's employment of these funds does not provide direct assistance to the short term interests of the vast majority of the bourgeoisie, as it is employed to guarantee the survival of the system as a whole.

Finally, a portion of these funds, which is at times considerable, is employed for services such as welfare payments, public health, public education, public housing, etc., that exclusively benefit the working class. Simultaneously, as O'Connor (1973) pointed out, the process of taxation is also resisted by the working class, which does not approve of State expenditures to subsidize capital accumulation. Nevertheless, the State is forced to face this resistance in order to create and maintain the general conditions for accumulation, i.e., it needs to legitimate. As it is necessary to maintain an adequate level of legitimation in order to create additional accumulation, and as legitimation is generally possible only if wealth is subtracted and removed from accumulation and reassigned to legitimation, the State must divert additional resources from accumulation to legitimation. The State soon finds itself in the difficult position of collecting the funds necessary to maintain the desired level of accumulation and legitimation, and/or having to justify ideologically its action. Both a fiscal crisis and a crisis of ideological legitimation is thus precipitated.[25] This contradiction is aggravated by the high level of difficulties encountered when plans are made to attempt to solve it. In reality, legitimation and accumulation are delegated to different social groups, accumulation to the bourgeoisie and legitimation to the officialdom of the State. These two groups pursue different and contradictory primary goals.

ACCUMULATION, LEGITIMATION AND SMALL FARMS

It is important at this point to link the general discussion on accumulation and legitimation to small farms and the small farm sector as a whole.

First, it must be stressed that the process of legitimation can appear under two principal forms, the material or economic and the ideological. These two forms can be analyzed separately, but their differentiation is only valid at the epistemological level. In reality they are indivisible, two sides of the same coin, and should be understood as such. The material aspect of legitimation can be defined as a material act which is relevant for both the reproduction of the system as well as the limitation of the contradictions within it.

An example of this form of legitimation in agriculture is the phenomenon known as the "Sponge Effect" (Mottura and Pugliese, 1980) (see Chapter 3), which refers to the process of limiting the surplus labor force (i.e., the unemployed) to one sector or area so that the chances that the contradictions present in society will rise to the surface are reduced. The Sponge Effect operated in Italy at the end of War World II (Fabiani, 1979; Graziani, 1979; Mottura and Pugliese, 1980). At that time the industrial apparatus underwent a rebuilding process that required relatively little labor. Unemployment was especially high in urban areas and in overcrowded rural areas that had experienced years of anti-urbanization measures implemented by the Fascist regime.[26] This unemployed rural population was very active, initiating many strikes and demonstrations in an attempt to find a solution to the situation. Such activities, which were organized jointly by peasants and farmworkers and supported by urban workers, created a series of social tensions that jeopardized the stability of the fragile democracy established after the war.

It should also be mentioned that communist beliefs were very popular among these workers so that the potential for a communist revolution represented a very serious threat to the ruling bourgeoisie. The solution, which the ruling class was forced to adopt, was that of creating through agrarian reform, a network of small farms that would occupy the unemployed rural labor force. This reform was a failure from the perspective of peasants and farm workers (Bandini, 1956). It did, however, prove to be extremely instrumental for the industrial development of the country, as it kept the labor force and, consequently, social pressure away from the cities. When industrial reconstruction was complete, this sector experienced unprecedented growth. Italy entered a period of prosperity that was greatly aided by the urban migration of peasant and farmworkers, who had discovered that the small farms could not provide adequate sustenance.

As it has been documented, the small farm sector in Italy became a tool to limit contradictions within the system. It provided a form of legitimation for the economy of the country. When industrial growth took place, a large number of small farms disappeared. During the same period, many policies in favor of small farms were revoked by the central government, largely due to the fact that the small farm sector no longer had to perform the role of sponge for the system.

Another and perhaps more familiar example of material legitimation is the case of small farms in the United States during the Great Depression of the 1930s, which has been illustrated in chapter 3. In that instance small farms provided subsistence possibilities to a large portion of the US population who would not otherwise have been able to survive.

The *ideological* aspect of legitimation refers to the creation of ideologies that are in one way or another supportive of the action of the ruling class (Gramsci, 1973; Poulantzas, 1978). A classic example of ideological legitimation was described in Chapter 2 when the events of 1848-1849 in France were illustrated. On that occasion the French peasantry was a determining factor in the rise to power of Napoleon III, and their support of the Second Empire was based on the ideological hegemony of the industrial bourgeoisie.

Another, and more modern, example of ideological legitimation is Agrarianism in the United States. As an ideology Agrarianism stresses the fundamental importance of small and/or family farms in the constitution and survival of American democracy and in the economic and moral prosperity of the country (Buttel and Flinn, 1975). Small independent farmers thus see themselves as one of the major elements constituting the essence of America, without considering the fact that their political influence has usually been weak. The concentration of land that took place in the United States after the Homestead Act (Faux, 1973) almost immediately limited the possibility of the small farms' independence. The technological revolution that followed subjected farmers to urban interests, and it then appeared evident that prosperity for farmers did not automatically mean prosperity for the rest of the population (Danbom, 1979). Additional, years of vocal support from the

Federal Government were only translated into realized support for the largest farms (Brewster et al., 1983; USDA, 1980; Vogeler, 1981). Agrarianism, therefore, lends an appearance of relevance to small family farmers in American society, while their actual relevance is much smaller. The potential for social tensions between small farmers and the ruling class is thus minimized. This action contributes to the establishment and maintenance of domination by the ruling class, which is by definition an act of legitimation.

A final example of ideological legitimation is that of "peasantization" in Italy (Mottura and Pugliese, 1975). As mentioned in Chapter 1, on two different occasions in the recent past the ruling class used the support of peasants to legitimize its power. During the 1930s the fascist regime obtained the support of the rural masses by assigning them small plots of land, which were not sufficient to guarantee any economic future to them but were enough to make them feel as though they were independent farmers.

Similarly, during the second post-war period the Christian Democratic Party (DC) obtained a large political consensus from the rural population after the implementation of land reform. The agrarian reform was not successful in the redistribution of land to peasants but did establish pro-DC feelings among rural masses, which still constituted one of the primary foundations of this party power.

The consideration that small farms can and do perform legitimative roles within capitalism does not make this activity an automatic prerogative of these agricultural units. Socio-historical conditions determine the degree in which accumulation and legitimation are present among small farms and which of these two roles is predominant (Mottura and Pugliese, 1975, 1980). It follows, then, that in order to assess the manner in which legitimation is delegated to the small farm sector in South Italy and Sicily, it is necessary to reconsider the position of the groups of farms identified above vis-a-vis the stability of the entire socio-economic system. This position, as previously identified, is determined by the attitudes of farm family members and by structural information related to the farm enterprise and the farm family members themselves.

Given these assumptions, the process of legitimation carried on by small farms in the regions considered takes several forms. Some of them can be classified as material forms in which legitimation is carried on, other as ideological forms.

First, all of the farm families classified in the Reserve group remain in agriculture due to a lack of job alternatives outside the sector. This pattern resembles that identified by Mottura and Pugliese as the "Sponge Effect," namely, the fact that small farms provide stability to the system by keeping a labor force in agriculture that, if released could produce serious social tensions. Farm income is the principal source of income for families in the Reserve group, and, despite the precarious socio-economic conditions in which they live and the rather low standard of living maintained, the farm enterprise is the only activity that can support them. Furthermore, as deprived agricultural workers, they are recipients of welfare payments, which contribute greatly to the family budget. It follows that for these family members the farm is the instrument through which their position as proletarianized and pauperized workers is partially concealed. That is, the farm and farm-related benefits prevent family members from actively and extensively searching for jobs in a labor market in which jobs are largely unavailable. However, due to the minimum economic support provided by the farm, their inability to find employment is not immediately translated into social unrest, as it occurred in other historical periods in which this condition was not present (Calza-Bini, 1976; Mottura and Pugliese, 1980).

It must be stressed that some differences exist between the farm families discussed in this study and those described by Mottura and Pugliese. More specifically, in the case of Southern Italian farmers in the fifties and sixties, the status of these "crystallized laborers" was a temporary one which ended when the economic development of the country generated a high and stable demand for unskilled labor (Mottura and Pugliese, 1980:182). Today, in South Italy the characteristic of crystallized labor assigned to these farmers assumes a more permanent posture, given the rather improbable availability of large amounts of unskilled or low skilled jobs in the near future (De Benedictis, 1980; Graziani, 1979). In other words the system's legitimative role as "sponge" tends to be an enduring characteristic of a portion of the small farm sector, rather than a transitory one.

Related to the form of legitimation just mentioned are the cases of farm families classified in the Complementary group. For families in the Complementary group, the farm represents a source of income that complements the principal off-farm earnings of the family. In this respect, the farm permits an adequate standard of living for the family, even though the family members' principal employment would not be itself guarantee these levels. The farm activity, then, legitimizes these family members' position as part-timers and makes it a constant and viable way of existence.

The farm activity also legitimizes alterations in the relationship between capital and labor in the off-farm labor market. In this case, the potential for tensions between labor and management are partially eliminated by the displacement of workers' demands from the realm of industrial relations. The workers' economic needs are not exclusively decided within the sphere of off-farm employment but are also delegated to the farm activity. Consequently, the real income of workers can be reduced without immediately generating social protest due to the fact that the marginal loss can be recuperated within the farm activity. That is, the farm allows the possibility for a reduction in the cost of labor by providing a supplement to the off-farm income that does not need to be negotiated directly by workers and management.[27]

The third and last form of material legitimation provided by the farm sector in South Italy and Sicily involves families classified in the Retired group. The persistence of these families in agriculture is related to the fact that their members have elected to spend their retirement years on the farm. The farm thus performs the role of keeper of a portion of the population for whom there is no immediate economic need in society (Bonanno and Calasanti, 1986; Riley, 1972), a role that may be viewed as an extension of the "sponge" effect performed by the farm sector for younger portions of the population. In this case, however, this legitimative role does not involve control of an active labor force but, rather, of a "permanently marginalized" one.

The selection of the farm as a retirement place has additional significance vis-a-vis the process of legitimation. It introduces us to the ideological aspects of legitimation that are delegated to the small farm sector. Among all of the families considered, this selection is based on an acceptance and support of traditional values associated with farming. As perhaps best illustrated in the case of farm families classified in the Traditional group, these values represent an extremely important condition for the persistence of these families in agriculture. More importantly, they shape the socio-political behavior of these individuals in the direction of traditional patterns of action.

The meanings of concepts such as "love toward the land" and "attachment to the way of life farming can provide," which are typical of these family members, display conservative attitudes toward society. For these families, as well as for others in various other groups, alteration of the status quo is generally viewed negatively in the case of both changes in the socio-political organization of the system and in the economic sphere. This allegiance to traditional values signifies a conformity to the dominant ideology in society, which consequently represents a legitimation of the social system. In fact, these values are translated into a set of attitudes and patterns of behavior that are supportive of the present social division of power and labor; that is, they are supportive of the hegemony of the ruling class. Furthermore, it is not a coincidence that the region surveyed is politically one of the strongholds of the Christian Democratic Party (DC), whose conservative policy appeals increasingly to rural southern residents while being strongly opposed in urban areas in the South and rural and urban areas in the North[28]. In view of their historical and cultural differences, the ideology of the farmers in question does not differ substantially from "Agrarianism" as found among American farmers. Conservative values and an appreciation of farm life represent the major ideological traits of these two groups, and in a similar fashion, both ideologies are integral parts of the hegemonic project of the ruling class in both countries.

As is clear from the evidence presented, small farms in the region considered hold a position within the social system in which their accumulative role is quite limited, if not null, and their legitimative role, quite high.[29] Viewed in these terms, it appears that the position of small farms in the system is ultimately supportive or functional to the reproduction of capitalism in Italy. However, as indicated in the previous unit, the relationship between accumulation and legitimation is a contradictory one that is reproduced at the sectorial level as the role of legitimation is delegated to social groups in spheres external to the State apparatus. In more concrete terms, this contradictory relationship in South Italy and Sicily takes the following forms.

First, the action of legitimation that farm family members perform by sharing conservative values and, consequently, supporting the existing dominant class (ideological legitimation) is contradicted by the precarious existence of the farms. The maintenance of an ideology stressing the desirability and superiority of farm life is thus undermined by the economic impracticability of these farms. Moreover, this ideology is eroded by both the limitations that the State faces subsidizing these farms as economic enterprises and by the diffusion of the urban values and ideology that largely dominate all Western countries.

Second, the financial assistance provided by the State to both the farm and farm residents cannot be maintained at current levels or increased without reducing the resistance of the bourgeoisie and various segments of the working class. The bourgeoisie resist taxation in general, but they are particularly opposed to the employment of tax funds for the support of segments of the population and productive sectors that are not related in the short or medium run to the process of accumulation. Specifically, in the last decade in Italy the organization representing agricultural entrepreneurs (*Confagricoltura*) has been increasingly vocal against State support of "inefficient" farms (Fabiani, 1979; Pugliese, 1983).

Portions of the working class resist State support to marginal farms as it implies a reduction in the amount of funds available to assist other segments of the pauperized working class. The existence of a large segment of unproductive agriculture also generates high food prices, which are opposed by the poor strata of the urban population. Consequently, the legitimation provided by the small farm sector is contradictory to social harmony in other segments of society. Furthermore, the State officialdom cannot afford to maintain such a large segment of unproductive agriculture without creating a food trade deficit that, in turn requires additional taxation of the country's population and a redistribution of State-managed resources among the various classes and class fractions. In both cases this operation is resisted by the bourgeoisie and the working class though for differing reasons.

During the past decade the increasingly unproductive nature of a large stratum of the small farm sector, especially in the South, created a food deficit that has grown at a rate of 1 trillion lira a year (see Chapter 3), making it the country's second largest trade deficit after that of oil products, of which Italy is a net importer. The Italian State's emphasis during the same period upon increasing accumulation among large farms (Fabiani, 1979, De Benedictis, 1980; Mottura and Puglises, 1980) testifies to its inability to halt the growth of this deficit and marks as ineffectual all attempts to correct the situation.

Third, and related to the above, the limitation of the process of accumulation to large farms signifies waste of resources both in natural and in human terms. Moreover, the lack of accumulation among small farms represents a negation of the very reason for existence of an economic enterprise. In Italy in the past decade, 3.5 million hectares of land once devoted to the cultivation of major crops were abandoned, and there was a 10 percent reduction in the total number of head of cattle. Of the abandoned land 2.5 million hectares were lost by small farms, while the same group of farms lost over 2 million head of cattle (Fabiani, 1979). The increase in the large farm sector could not compensate for these losses, seriously limiting the ability of Italian agriculture to generate an annual production sufficient to feed the country's population. At the occupational level the loss of resources has been paralleled by a loss of jobs in the sector. A relevant portion of the lost jobs, however, were not replaced by others in non-farm sectors, as industry and the tertiary sector experienced a prolonged crisis (Graziani, 1979a; Pugliese, 1983). Rather, there was[30] an increase in part-time activities and in unemployment. The situation of several farm families in the Traditional and Reserve groups are all cases in point.

Finally, and more importantly, lack of accumulation and the role of accumulation performed by small farms are contradictory to alterations in the dualistic pattern of agricultural development that has characterized Italy during the past three decades (see Chapters II and III). In fact, as long as the small farm sector in South Italy and Sicily is assigned the role of keeper of surplus population (Sponge Effect), no major changes in the present productive conditions of these farms will be possible. Small farms can be relieved of their role as sponge of the system only if new jobs are generated for the underemployed farm labor force.

This solution is at present highly impracticable for two major reasons. First the crisis of the non-farming sectors does not allow for a large occupational expansion, especially in poor regions like the South. Second, the creation of new full-time jobs in non-farming sectors contradicts the process of industrial decentralization (Calza-Bini, 1976), which is based on the availability of cheap labor (often hired on a part-time basis) and has been one of the most effective entrepreneurial responses to the rising costs of production.

From another point of view, an improvement in the productive and, consequently, accumulative potential for small farms is not likely to generate more farm jobs but, rather to further expose underemployment and undermine the very legitimative role performed by these farms. In other words, the implementation of dominant agricultural policies aimed at the increase of production and productivity would immediately result in, among other things, an increase in mechanization and an expulsion of the labor force, which would be in open contradiction to the need for new jobs. Furthermore, this last solution would involve the employment of State financial aid exclusively for small farms, a policy that has been long abandoned and would be, and is, opposed by dominant groups both domestically and at the EEC level.[31]

The concentration of the process of accumulation among large farms has widened the productive gap between the large and small farm sector. This situation is largely the outcome of an agrarian policy characterized by the overt assumption that it would benefit small farms as well as, or even more than, large farms and by results that largely penalized the former group of farms (see Chapters 2 and 3). Consequently, the legitimative role performed by small farms is jeopardized by the discontent generated by the results of this policy. In order to counter this legitimation crisis, the State needs to strengthen further the sources of allegiance of small farmers to the system. In other words, the State needs to reinforce the ideology of agrarianism among small farmers, a practice that cannot be carried out without creating further contradictions.

The State, however, can pursue material forms of legitimation as an alternative. That is, it can attempt to devote more State managed funds to assist small farmers. As noted earlier, serious problems for the implementation of this strategy also develop. Moreover, the State is forced into a position in which it must divert resources from accumulation to legitimation in order to stimulate accumulation. But the more it tends to stimulate accumulation, the more it needs to reinforce legitimation. The more the State needs to channel capital toward accumulation, the more it needs to use the same capital for legitimation. The State, then, finds itself in a contradictory situation in which it cannot effectively balance accumulation and legitimation. Yet, accumulation and legitimation, as well as capitalism, cannot be balanced automatically by the "invisible hand" of the system. Consequently, the State's inability to perform the role of balancer of the system is immediately translated into the necessity to maintain some degree of disequilibrium within the system itself. In this specific case, it is translated into the continuous existence of a dualistic pattern of agricultural development.

It follows that the existence of a dualistic structure of agriculture in Italy takes a permanent posture rather than a transitory one, which allows the continuous persistence of small farms. This persistence, then, is the outcome of a distorted process of development and not merely the ability of that agricultural enterprises to compete on the market.[32]

To conclude, it must be noted that the permanent character of the existence of small farms and the inability of the capitalist State to resolve contradictions do not necessarily have to lead to the assumption that improvements, both in the situation of small farms and in their positions within the system, are impossible. Based on the contradictory nature of the relationship between classes in capitalism and on the contradictory motion through which accumulation is reproduced, it is possible to hypothesize new directions of development. However, these directions cannot be fostered, as several Marxist theorists think, in a theoretical framework in which predetermined laws of capitalist development generate and control historical situations within the system. Nor can they be hypothesized as a consequence of an unavoidable collapse of the capitalist system under the unbearable weight of its contradictions. Rather, it is through an understanding of these contradictions as generated by class conflict and the critical analysis of their implications in society that new and more equitable forms of development can be found.

The objective of the next chapter is to examine the implications and perspectives that the analysis developed for Italy has in regard to the case of the United States. The availability of a number of studies and official data on both the American agricultural structure and the relationship between farming and the other socio-economic sectors allows us to identify trends, despite the absence of primary data such as those examined above. In the case of the United States emphasis will be placed on the implications that the considerations made for the case of Southern Italy have on the US, rather than that on the reasons for persistence as such. The first step in this direction will be to analyze further the relevance of the role of the State in advanced Western societies.

7
The Persistence of Small Farms in the United States: Implications and Perspectives

INTRODUCTORY CONSIDERATIONS ON THE CRISIS OF THE STATE AND ITS RELEVANCE IN ADVANCED WESTERN SOCIETIES

The discussion in previous pages has revealed the importance of the role that the State plays in advanced Western societies today. More importantly, it is evident that an analysis of the relationship between social actors and the socio-economic structure in mature capitalism cannot be fully explored without addressing this issue. Historically, the State's role has become more apparent through its continuous and growing intervention in spheres that traditionally were outside its competence, a trend largely due to its adoption of Keynesian political economic strategies. In the case of the United States such strategies have been a constant since the period of the Great Depression and the era of the New Deal.

The role of the State in all advanced Western societies since the end of World War II has been that of general mediator between the conflicting interests of the major social classes. This role of "pacifier", as Offe (1984) has illustrated, consists first (but not primarily) of the support of important needs of the working class. In more specific terms, through social services the State provides assistance to members of the working class who are marginalized by the very effect of competition in a market society (unemployment, low wages etc.). Simultaneously, the State regulates disputes between the working class and the capitalist class through legislation and direct arbitration. This normative role of the State involves the legitimaion of the role of trade unions (and other organizations of the working class, such as political parties) in society and, through this, the limitation of the range of action of the working class against the bourgeoisie. Second, the State guarantee the possibilities of accumulation and, consequently, supports the interests of the bourgeoisie. However, as illustrated in Chapter 6, these interests do not correspond to the short term needs of any specific fraction of the bourgeoisie itself, but represent the interests of the class as a whole. The State is thus supportive of the general conditions that allow long term accumulation in society. From this point of view it is possible to argue, as we did above, that the position of the State as mediator is not a neutral position, but a class position. It favors the rule and leadership of the bourgeoisie and the continuous existence of the present relations of production (Block, 1977; Offe, 1980; Poulantzas, 1978).

In the course of the last decade the role of the State as "pacifier" became increasingly untenable as fundamental contradictions emerged in mature capitalist societies. As has been documented for Italy, the State was unable to solve the problem of regional underdevelopment or to improve the socio-economic conditions of the working class and a substantial number of members of small farm families. This inability of the State to address the needs of the working class exposes the State itself to a mounting legitimation crisis to which it is unable to respond without compromising possibilities for further accumulation. In this respect the legitimation crisis is accompanied by an accumulation crisis that generates tension between the State and the various segments of the bourgeoisie. More importantly, the accumulation crisis limits the autonomy of the State and its range of action as its potential for acquiring financial resources is eroded (Offe and Ronge, 1979).

The problem of accumulation crisis has been a constant concern for the bourgeoisie as a whole in recent years, and it has been addressed in terms of alternative political strategies in the economic sphere and in the normative sphere of State intervention (Offe, 1984). These strategies, which have been associated with conservative political positions, demand a return to a laisser-faire posture in the economic realm and plan measures to increase labor productivity. What is argued is that the State, through its intervention, has created a number of restrictions on the possibility of maneuvering of the bourgeoisie, both in terms of taxation and regulations. These restrictions constitute de facto disincentives to investments. Simultaneously, State action in support of the working class represents another limit on the bourgeoisie in its effort to increase labor productivity. Regulations and labor legislation do not allow an increase in workers' productivity as would occur in free market situations. The State is thus held responsible, from this political point of view, of promoting among workers a "disincentive to work" (Offe, 1984, Giori, 1983).

The historical inability of the State to perform the role of "pacifier" in mature capitalist societies is also acknowledged by non conservativ groupings. In the progressive camp it has been emphasized (Offe, 1984) that the State today is "ineffective and inefficient." As far as welfare policies are concerned, the State has been charged with adopting an ex post facto strategy. That is, the State's employment of a welfare policy does not eliminate the sources of marginalization and exploitation of the working class, but intervenes only to limit their consequences. Simultaneously, though the State provides protection to the working class through labor legislation and welfare, it legitimizes exploitation. As Offe (1984) would argue, members of society in need of State assistance not only must prove that they are in need but must also to prove that they deserve it. They must clearly manifest conformity to values, beliefs, cultural and economic standards of the society.

Alternatives to the "ineffectiveness and inefficiency" of the State have been provided from both the conservative and the progressive camp. However, as is apparent from the Italian case as well as from other analyses (Offe, 1984; Poulantzas, 1978), these solutions present a number of problems.

Perhaps, the most debated proposal of all is that of the elimination of State intervention in spheres that are external to its original normative competence. More specifically, a withdrawal of State action from the economy, in terms of both production and labor, and from the civil society in terms of reduction and/or elimination of welfare programs has been proposed, so that a laisser-faire society would be generated.[33]

As with the case study of Italy illustrated above, recent studies (Offe, 1984; Poulantzas, 1978) suggest the inapplicability of this solution in its integrity. It is increasingly evident that the proposal to return to laissez-faire strategies are limited to only a few segments of the area of State intervention. In other words, the proponents of this political approach are inclined to support the idea of disengagement of the State in spheres such as welfare programs and regulation of business activities, but are not willing to renounce its support in other areas that directly or indirectly pertain to their own economic interests. As Offe (1984) points out, the bourgeoisie in advanced Western societies is reluctant to do without all the benefits that it derives from the action of the State. To use O'Connor's (1973) expression, the bourgeoisie is unwilling to renounce "social capital expenditures" and "social expenses of production;" it is unwilling to relinquish subsidies for private capital accumulation and to have the private sector cover the social costs of private production.

The case of Italy provides further evidence in this direction. As has been illustrated, the Italian bourgeoisie cannot afford to have the State interrupt its welfare farm programs, either at the family or farm level. The specific motives for this reluctance lie not only in the sphere of legitimation but, above all, in that of accumulation. The process of decentralization and informalization of economic activities in the Italian case occupies a relevant position in the country's expansion of the forces of production. As has been mentioned estimates of the so-called underground economy place its contribution at over 20 percent of the GNP. This phenomenon signifies that, not only accumulation, but also social consumption can take place. Social consumption, as pointed out by Mingione (1981a), is fundamental in mature capitalist societies, as crisis assumes the characteristics of overproduction (or, as viewed from a symmetrical point of view, of underconsumption). Within the Italian economic system, then, the intervention of the State in terms of welfare programs is a stimulus to consumption and thus indirectly to production. If the Italian bourgeoisie had to limit State support exclusively to the productive sphere, it would accelerate the crisis of overproduction/underconsumption. From a different point of view, however, the maintenance of a diffuse yet economically inadequate welfare system forces many families to remain in agriculture and/or to search for low paying marginal jobs outside the sector. This phenomenon, though the foundation of a process of accumulation, is also one of the principal factors responsible for stagnation and underdevelopment in many regions of that country. The Italian case, then, indicates that neo-laizzez-faire strategies could be applied only partially and not without contradictions.

Offe (1984) provides a summary statement of the fundamental contradiction in advanced Western societies. While capitalism cannot coexist with a State which intervenes in economic and social spheres, capitalism itself cannot exist without a State as it is known today. Alternatives that can partially limit the magnitude of this contradiction are possible. In particular, an application of laissez-faire strategies can be successful in the short run.

In the United States there has been a clear attempt in recent years to introduce neo laissez-faire policies (Devine, 1985). The rationale for the introduction of these strategies is largely based on the inability of the State to overcome a decrease in the rate of profit vis-a-vis its fiscal crisis (Green and Carlin, 1985).

This attempt, however, displays the same contradictions illustrated above. While measures aimed at the reduction of welfare spending have been introduced, there has also been a continuous demand on the part of the bourgeoisie for the State to be engaged in operations aimed at sustaining accumulation (Devine, 1985; Castells, 1980).

The American State is confronted, however, with a relatively different situation than that of its European counterparts. This phenomenon can perhaps be attributed to two diverse yet intimately related factors: the different magnitude of State intervention and the level of accumulation on the two continents. As has been documented (Giori, 1983; Offe, 1984; Orloff and Skocpol, 1984), in the United States the intervention of the State in the economy and civil society is less developed that in the so-called "Social Democracies" of Western Europe, such as Great Britain, France, Germany, Italy, the Low Countries and the Scandinavian Countries. At the same time, in recent years accumulation has been higher in the United States than in Europe. As illustrated by an analysis of basic economic indicators, the expansion of the forces of production in the US have not been matched by that of any of the Western European Countries. Since the beginning of this decade the average growth rate in the U.S. has been four percent, coupled with an inflation of less than five percent (USDC, 1985). Western European Countries never reached expansion rates of that magnitude during the same period. The average rate of growth for Western Europe during the eighties has been 1.5 percent and inflation has been well above the

American value, in many cases in the double digit figure (USDC, 1985).

Given the differences between Europe and the U.S. and the differences in the relationship between the State and the socio-economic sphere between the two continents, what are the implications that can be drawn from the study of the Italian case for the agricultural sector and, in particular, vis-a-vis the persistence of small farms? Furthermore, from a micro-analytic point of view, what relevance do the conclusions reached for Italy have for the persistence of small farms in the U.S.?

In the following sections an attempt to address briefly these issues will be made, yet without any pretension of providing a complete explanation to them. Rather, the arguments are to be taken as hypotheses to be evaluated further in the future. Ultimately, they could constitute important research topics in the realm of the sociology of agriculture.

PERSISTENCE AND FARM FAMILY MEMBERS' PERCEPTION

In previous chapters of this book the persistence of small farms in Italy has been approached from both the micro and macro levels. At the micro level farm family members' accounts of their work in agriculture and the persistence of their farms have been considered. At the macro level the structural and societal factors that contribute to the persistence of these farms have been analyzed. In this unit it is our task to address briefly the persistence of small farms in the U.S. from a micro point of view.

The analysis of small farms in Southern Italy has made it clear that a large portion of small farm family members perceive their participation in farming as not related primarily to economic factors. Emotional, cultural and ideological factors are considered their principal motives for remaining on the farm, though they acknowledge the importance of farm income and other economic motives.

The largest group of farm families among the six identified is composed of families who remain on the farm because of emotional attachment to the land and/or farm life. As has been illustrated above (Chapter V), the emotional attachment to the land and to the way of life that farming represents can overcome the hardships of insufficient economic rewards associated with farming.

Similarly, other categories of farm families base their motivations for remaining in agriculture on non-economic factors. For example, though the members of families belonging to the Complementary group, stress the importance of farm income as a complementary factor in their total family income, they also emphasize the importance of non-economic factors for the continuation of their farming activity. They receive true satisfaction from living on the farm and participating in farm life. Given the very limited amount of family labor required on the farm, farming is considered more a way of life than an occupation.

The Retired group represent another case in which non-economic factors are predominant in these families' continuation of the farm activity. For the individuals included in this group farming is the activity that they selected for their retirement years. They emphasize the enjoyment and relaxation that farm life can provide them. Similarly, families included in the residual group, such as those for whom farming is a hobby, can be considered in this larger group of people who view farming primarily as a non-economic activity.

In the United States there are indications that attitudes favoring the continuation of farming for non-economic reasons exist among small farm family members and that these attitudes are perhaps more wide spread than in Italy. As part-time farming has been largely associated with small scale farming both in the U.S. and abroad (Cavazzani, 1980; Coughenour and Gabbard, 1977; Fuguitt et al., 1977; Schroeder et al., 1985), the literature on part-time farmers' attitudes toward farming can be employed to support our claim. In his study of part-time farmers in Kentucky, Coughenour (1977) indicated that social, cultural and emotional rewards attached to farming are as important, if not more so, than economic benefits associated with the farm activity. He writes: "Farming as a business venture is not the only, and for many perhaps not the primary, interest that they have in farming" (1977:i). Similarly, comparative studies of U.S. and European part-time farming (Arkleton, 1983) point in the same direction: "Side by side with ... economic factors, what may be called 'ideological' or 'non-materialistic' influences have also reinforced the turnaround in rural depopulation and the endurance and expansion of part-time farming" (Arkleton, 1983:35).

Other segments of the literature on small scale farms also provide evidence in favor of our thesis. In his study on farm income Crecink (1979) stresses the relatively low importance that farm income has for small farm families. Similarly, a recent USDA publication indicates that "farmers depend heavily upon nonfarm sources of income. Over three fifths of the farm population's personal income came from nonfarm sources in 1982, compared with only two fifths in 1973" (USDA, 1984). Furthermore, the proportion of off-farm income out of the total family income increases with the decrease in farm size. On the average, in past years families operating farms with at least $40,000 worth of sales received 30 percent of their income from off-farm sources. For families operating farms with lower total sales, the off-farm income is usually larger than the farm-income and in many cases offsets losses in the farm operation. In the 80's the off-farm income for American small farms has averaged 103 percent of the total income, as the latter is burdened with negative farm income (USDA, 1984a).

These studies, while not addressing directly the importance on non-economic motivations for farming, implicitly indicate that income related factors are not the most important reasons for the persistence of American small farmers in agriculture. These phenomena also suggest that small farm families in the U.S. view farm work more as a lifestyle than an occupational role. This is the hypothesis presented in a recent study (Schroeder et al., 1985) in which indicators explaining the economic "anomaly" of the continuous existence of small scale farmers are analyzed. According to this study, if farming is presented in terms of maximization of profit, the persistence of small farms can only be treated as an unusual event. In fact, "small scale farms rarely have the resource base either to optimize efficiently or to provide sufficient income for the farm household. Yet, small scale farms are a tenacious and possibly even growing segment of the farm population" (Schroeder et al., 1985:319). According to these authors, a possible solution is to investigate the importance of lifestyle and its relationship to economic factors in assessing the phenomenon on the persistence of small farms. Moreover, it is argued that lifestyle variables are not only relevant to the explanation of the persistence of small farms in agriculture, but as the farm economic crisis expands, they become relevant to the decision of families operating larger farms to remain in the sector.

A similar position has been argued by Kliebenstein at al., (1981). They write: "A principal assumption behind much farm management research is that a farmers' basic goal is to maximize the difference between costs and profit. Maximizing the profit is an important force in farm level decision but, as shown in this study it is not the only force influencing decisions." (Kliebenstine, et al.,

1981:10) The alternative forces that the authors refer to are cultural and ideological variables. More specifically, a sample of Missouri farmers has shown that for them the most important reasons for continuing farming are those of "doing something worthwhile" and "being its own boss." In other words non-economic factors predominant over economic factors in motivating the continuation of these families' farm activity. It is also emphasized that ideological, cultural and emotional factors are more important among farmers whose operations did not show a high rate of growth. The reverse, however, is only partially true, as farmers whose operations recorded a faster rate of growth value both economic and non-economic factors as their principal motivations for remaining in farming. The authors conclude: "Thus, for the group of farmers experiencing the most rapid growth, maximizing income was not their only major goal as they valued 'being own boss' as the most important item. However, it does point out that those that felt they were more motivated by income did in fact have a faster growth rate" (Kliebenstine et al., 1981:13).

The importance of ideological, cultural and emotional factors in the phenomenon of persistence in the United States corroborates the explanation provided by Max Weber. As has been illustrated in previous chapters (2 and 5), for Weber the persistence of small farms in agriculture, despite unfavorable market conditions, is tied to the freedom and independence that this activity carries with it. Evidence for the United States (Kliebenstine et al., 1981; Schroeder et al., 1985) shows that these attitudes permeate all strata of the farming spectrum and by no means are confined to small farms, though small farms families' members tend to pay more attention to them.

The Weberian theory also accounts for the selection of non-profit maximizing strategies among farmers and is supported by an increasingly large body of literature on farming in the U.S. that points to alternative farm management strategies (see Smith and Capstick, 1976; Swanson and Bonanno, 1986).

The assumption that farmers tend to select profit maximizing strategies has become increasingly untenable under the present farm crisis and the structural changes that have accompanied it. In a recent USDA publication (1985) on the debt/asset ratio, it is suggested that farms which have been managed with a profit maximization orientation tend to have a higher debt/asset ratio than others that chose non-profit maximizing strategies. This, in turn, signifies that the latter type of farm has more of a chance to persist in agriculture than the former, which is largely in debt. Furthermore, a substantial number of small farms in the U.S. have used non-profit maximizing strategies and, consequently, according to the logic of the analysis of the USDA, tend to have been a greater possibility of remaining in agriculture than other farms.

Related considerations can be made for "new entries" in the small farm sector. It has been pointed out (Chantfort, 1982; Brooks, 1985) that most new small farmers do not farm by necessity but by choice. For many of them, "farming is only a part-time, extra-income activity often chosen more as a lifestyle than an occupation" (Chantfort, 1982:4). Consequently, and given the high risk that investing in a commercial operation carries with it, it is safe to assume that for most of these farms the preferred alternative is that of non-profit maximization strategies.

The accuracy of the statement of USDA officials is to be determined by future events. In this context it is relevant to reaffirm the importance of cultural, ideological and emotional reasons that motivate small farm family members to continue their work in farming. This tendency is strong both in Italy and in the United States, making it a common characteristic of both groups of farmers. It also opens new avenues for the theoretical reading of farmers' behavior and the construction of socio-economic alternatives for this segment of the population.

This is not to say, however, that economic factors are not important in the overall motivations of farmers to remain in agriculture. As indicated above (chapter 5), there is no single denominator that can account by itself for the persistence of small farms. Rather, a complex pattern of elements, both in the economic and non-economic realm, can better account for the phenomenon in question. In the case of Italy the overall condition of the economy makes farming the only viable chance that specific groups (such as the Reserve group) have to be engaged in gainful work. In the U.S. this type of alternative seems to be less important. First, the overall condition of the U.S. economy (higher rate of growth, lower inflation, lower unemployment rate, smaller proportion of workers engaging in farming, etc.) provides more job opportunities than those available for the Italian labor force. Second, the growth of rural off-farm job opportunities in the U.S. has not been experienced in Italy, so that the Italian rural labor force has to compete for fewer available off-farm jobs. Finally, small farms in the U.S. tend to have negative farm incomes, so that income is more a deterrent from farming than an attractive feature. However, regional socio-economic characteristics within the U.S. may alter this picture to resemble more closely the case of Italy.

Economic factors that influence the farm family members' decision to remain in agriculture are, by no means, confined to the farm income. They may also include items such as off-farm employment opportunities, off-farm income, labor market conditions and economic processes occurring in other socio-economic sectors. The case of Italy provides ample evidence in this direction, as the decentralization and informalization of industrial activities to rural areas, the economic role played by the State and the status of local rural labor markets influence the persistence of small farms and family members' perception of their role in farming. In other words, despite the fact that farm income is insufficient to support the farm family and is considered a secondary factor among the reasons for persistence, the presence of these farms and of the labor associated with them set in motion a series of processes centered around economic elements.

PERSISTENCE: A STRUCTURAL VIEW

The considerations made above relate directly to the Kautskyan theory for the persistence of small farms. This theory (see chapters 1 and 5) stresses that the persistence of small farms is linked to the role of producer of surplus labor that these farms perform within the socio-economic system. Furthermore, it provides insights into the relationship between petty commodity production and accumulation, indicating that the persistence of a productive unit within capitalism is not necessarily related to its ability to produce nor to the productivity or production of the unit itself. Rather, there are instances in which the productive role of agricultural units is second to other "social" roles whether in the implicit or explicit form. Furthermore, in formulating his thesis Kautsky extended the Marxian notion of the role of agriculture in primitive accumulation to a historical context in which primitive accumulation has been reached. In the Marxian theory of primitive accumulation the role of agriculture is that of producer and dispenser of labor to be utilized in manufacture (see Marx and Engels, Capital, chapter XXIII and XXIV). The urbanization of agricultural labor (peasants) is, in fact, a fundamental condition for the expansion of manufacture and the development of the process of accumulation. Kautsky reformulates this notion in a context of advanced capitalist relations where the flow of labor from agriculture to manufacture is not always continuous and in which agriculture itself has reached a substantial level of capitalization.

The monolithic tone of his observations, however, limits the power of his original explanation (see chapter 1). Yet, if reinterpreted, it assumes fundamental importance. For the case of Italy, several reinterpretations and reutilizations of the Kautskyian formulation have been provided. Among these, the versions of Calza-Bini (1974;1976) and Mottura and Pugliese (1975;1980) discussed in the introductory chapters assume particular importance.

For Calza-Bini the persistence of small farms in marginal regions of Italy is linked to the proletarianization of farm labor and the parallel process of industrial decentralization. Decentralized industrial units, which require labor for low-paying jobs, can count for their expansion on the displaced farm-laborer who lacks alternatives. According to this author, these workers are proletarianized because their farming activity goes through a three stage process which consists of a reduction in farm income, production in loss and, finally, subsistence production. At this point family farm members are forced to sell their labor outside the farm. For Calza-Bini, as for Kautsky, small farms produce proletarianized labor to be utilized off the farm, but for the former this labor is mostly utilized by decentralized industrial activities.

For Mottura and Pugliese the case of persistence of small farms refers primarily to the role of keeper of surplus labor. Accordingly, small farms tend to persist (and in periods of general economic crisis even to increase in relative terms) due to the fact that they perform the role of "sponge of labor for the system." They retain or release the labor force according to the general trends of the domestic and international markets and the fluctuation of the labor demand. In periods of recession, which are characterized by a higher level of unemployment, small farms keep labor crystallized in agriculture. When periods of expansion occur, labor is freed and made available for other economic sectors (industry and tertiary). In this theoretical framework, unlike that hypothesized by Kautsky, the small farm sector does not generate surplus labor, but retains it.

Both of these reinterpretations of the Kautskyan theory are important in analyzing the Italian case. By focusing on the overall structural trends taking place within the society, they also emphasize economic aspects which do not appear to be very relevant at the micro level. In this respect, the generation and utilization of family income assumes a central position in the analysis of the issue in question. Two alternatives are relevant for discussion. First, off-farm income is employed to supplement farm income. However, the gross farm income is low in relation to the costs of production so that off-farm income is employed to finance the farm operations. Off-farm employment is relatively stable within the farm family so as to guarantee a continuous flow of financial resources to the family in itself. Farm income is not necessary for the survival of the family. Second, the off-farm income is only partially sufficient to supplement farm income and farm costs of production. There is, then a decrease in the family's standards of living to adjust to the inadequacy of off-farm and farm income. Off-farm income is generated by precarious low paying jobs, and farming is a necessity rather than a choice. In the interpretation of the Italian case the second alternative is considered the general framework for interpretation.

What is, then, the importance of these considerations for the United States? The historical conditions of the U.S. make both alternatives relevant. It has already been documented that a large number of small farm family members have off-farm earnings that supplement an often negative farm income. Nevertheless, processes similar to those occurring in Italy have been observed in the U.S. Among these, decentralization of industrial production, informalization of production and the farm as "keeper of surplus labor" are of particular importance.

The notion of decentralization as employed for the case of Italy refers to the process of reorganization of production into smaller units in non-industrialized areas. Productive activities once executed in larger industrial plants located in heavily industrialized regions are relocated to rural and/or marginal areas, often under the form of contracting. Moreover, in Italy the process of decentralization has been associated with the process of informalization. The latter refers to the adoption of production strategies that employ unconventional industrial arrangements that range from cottage industry to piece work and from the use of illegal labor to tax evasion. In more specific terms, a substantial segment of the Italian industrial network has experienced difficulties due to the rising costs of raw material, energy and labor. These enterprises have been faced with the need to select alternative strategies to avoid a furthering of the crisis. Given the inelasticity of the labor market in the primary sector, the solution was found in delegating production to units in the secondary labor market which could be more flexible in their arrangement of production techniques. The large firm, however, remains in control of the product as it provides in most cases the vast majority of the inputs necessary (monopoly side) and purchases the final product (monopsony side).

As indicated above (chapter 4), the process of decentralization in Italy was a spontaneous one, for previous State organized efforts to locate industrial plants in marginal regions had largely failed. During the 1950's and 1960's the State attempted to industrialize the largely rural South through the use of infrastructure strategies, poles of development and financial aid to corporations willing to relocate. The result was the creation of industrial plants that did not generate development or enough jobs. In Italy decentralization of activities by branches of corporations operating in the core sector signifies the establishment in the periphery of the same wages and fringe benefits received by workers in core regions. A strong trade union movement, which was able to eliminate compensation differences among workers (within the same industry and with similar tasks) independently of the situation in the labor market, was perhaps the major factor in shaping this outcome. The situation left room for a decentralization connected with an informalization process and for the utilization of farm labor in marginal low-paying off-farm activities.

In the United States the process of industrial decentralization presents both similarities and differences with respect to the case of Italy. First, it was generated by the same crisis of profit experienced by Italian entrepreneurs. The increase in production costs and the strength of the labor in the core sector provided ample support for decentralization strategies. Accordingly, "industry has been rapidly decentralizing over the past two decades" (Chantford, 1983:9), both domestically and internationally. At the domestic level (which is our focus) decentralization meant the relocation of industrial activities to rural areas in primarily Southern and Western regions (Chantford, 1983; Horan and Tolbert, 1984; Martinez, 1985; Swanson and Skees, 1985).

These industries are generally low-wage, labor intensive manufacturing establishments such as textile mills and plants producing clothing, furniture, leather goods, rubber and plastic products and also in some cases automobile plants. Such industries can greatly profit from the cheap labor and low costs of production available in rural areas. Contrary to popular belief and unlike the case of Italy, these establishments tend to be branches of corporate firms, and a large segment (80 percent) of rural employment tends to be provided by establishments with more than 100 employees (Chantford, 1983; Martinez, 1985). Furthermore, in many cases plants in rural areas tend to be larger than similar ones found in old industrialized regions. "One factor (for the occurrence of this phenomenon) is lower nonmetro land costs, which encourage sprawling complexes employing many workers. Another is that large, mature firms which often choose to settle in nonmetro establishments are large simply because they are not surrounded by specialized suppliers, but have to produce many of their own inputs" (Chantford, 1983:9).

The decentralization process found in the United States also differs from the Italian case with respect to the role of the State. While in Italy State organized efforts were largely unconsequential, in the United States a great deal of emphasis has been placed by local governments and agencies on attracting firms to rural areas. Special financial arrangements (such as loans at lower interest rates or grants), commitments to partially or, in time, totally sponsor the construction of the plant, and/or commitments for the development of infrastructure pertinent to the operation of the plant are often provided to corporations that are willing to invest in the region. These strategies, coupled with the employment of cheap and non-unionized labor, have been particularly successful as the process of decentralization continues.

Despite the creation of new jobs, the decentralization process has been associated with rather unstable employment. These jobs are often exposed to foreign competitions as non-domestic producers can count on even cheaper costs of production. It has been stated that: "In the past, these industries have been shifting from metro to nonmetro areas, providing an important source of employment growth in rural and small-town American. Now, however, an increasing shift of these operations overseas jeopardizes this source of jobs. Because the low-wage, labor-intensive industries employ a relatively large proportion of the local labor force in some rural communities, the loss of such establishment can hit these communities especially hard" (Chantfort, 1983:9).

In relation with this industrial decentralization, an understanding of the farm and farm activity as a limited, yet secure, source of income and shelter for farm family members can be considered a very plausible hypothesis.

Informalization[34] is the other process which has played an important role in the persistence of small farms in Italy. Though it has also been studied in the United States, a great deal of attention has been paid to the phenomenon in urban, rather than rural, areas (Castells, 1984; Portes and Stepick, 1985; Sassen Koob, 1984). Nevertheless, there is evidence that informalization occurs in rural areas as well. In their investigation of the crisis of Midwestern farms, Heffernan and Heffernan (1985) found that farm family members are involved in informal activities. They indicate, for instance, that the production of high school yearbooks relies heavily on the domestic work of rural housewives in Central Missouri. These women receive a word processor, manuscripts to be typed in formats ready to be printed and are paid by the key hit. Similarly, it is common practice among rural residents in the Ozark region to be engaged in part-time activities in the tourist industry without the required licenses or permits or to be informally employed in the local shoe industry. Perhaps a more familiar case, often reported in the media,[35] that of the knitting homework activities in the Eastern portion of the country. Knitting entrepreneurs provide local residents (mostly housewives) with machinery and raw material. They then commission a final product and pay by the piece. In this manner produc-tion costs are reduced to a minimum and labor disputes are virtually eliminated, while for the house worker this activity represents a complement to the principle income.

Though more investigation is needed, these and other cases indicate that a supply of labor available for marginal activities exists in various rural regions of the United States. Moreover, in many instances this labor is the same as found on small and/or part-time farms.

It should be emphasized that the role of keeper of surplus labor performed by the small farm sector is not extraneous to the case of the United States, as examples can be found in the recent history of this country. A now classic case is that of the Great Depression. As illustrated above (chapter 2), during that period there was an increase in the number of small farms due to a process of migration to rural areas. The economic crisis in urban areas left workers no option but to return to the farm from which they had migrated a few years earlier. A similar phenomenon occurred in some marginal areas in more recent years. As documented by White (1983), in-migration to marginal rural areas such as the Appalachian region during the late seventies and early eighties may be viewed as an attempt to weather the economic crisis. As the crisis affected the industrial regions which are historically the destination of rural migration (North and Central States), former rural residents returned to their places of origin to some type of work on the farm in many cases. This return occurred despite the lack of economic improvement in marginal regions so that once new job opportunities developed elsewhere, a new process of rural out-migration was triggered.

The role of keeper of surplus labor is not new to the small farm sectors. Actually, it has been a fundamental characteristic of this type of farm in advanced societies since the period of primitive accumulation. The importance of this process at the present time is that it guarantees a safety valve for the system in the event of a large surplus of labor on the market. Both in the case of Italy and in that of the US the role of keeper of surplus labor performed by small farms has achieved this result.

The nature of the processes discussed above allow for contradictions, as demonstrated by the case of Italy. The magnitude of these contradictions in the United States vis-a-vis the historical conditions of the country needs to be assessed through a careful study of the problem. Nevertheless, their characteristics can by hypothesized based on the discussion presented above. In this case, then, some of the conclusions reached for the case of Italy can be reevaluated for the United States and, more specifically, for the case of marginal rural regions where the phenomena of decentralization, informalization and "keeper of surplus labor" assume particular importance.

SOCIO-ECONOMIC CONTRADICTIONS, REGIONAL UNDERDEVELOPMENT AND THE STATE: AN AMERICAN DILEMMA?

The issue of the persistence of small farms in the United States and Italy reveals a dual and overlapping feature. If the problem is approached from the micro point of view, in both countries emotional, cultural and ideological factors become fundamental in explaining the phenomenon in question. However, when macro analysis is attempted the relevance of economic factors is clear. Economic and non-economic factors thus play important roles in the explanation of the phenomenon, depending upon the angle of analysis. In more specific terms, rather than being mutually exclusive the two sets of factors, are complementary elements of a complex pattern so that both are sides of the same coin.

In the case of the United States emotional, cultural and ideological factors constitute strong and stable characteristics of farm family members' view on the issue of persistence. Conversely, the availability of off-farm jobs and farm income are often of a rather precarious nature. Furthermore, the potential for a change in the level of precariousness of these factors is increasingly related to the action of the State. In the case of off-farm industrial jobs, for instance, the action of the State is fundamental in attracting corporations to rural areal. State action is also central in maintaining these jobs in loci given the fluctuations of the markets for these commodities and the mobile nature of the capital involved. Even fluctuations of moderate magnitude in costs of production or in marked sensitivity to the supply can generate, as indicated by recent economic history, shifts in the location of production plants. The role of the State is perhaps even more important with regard to farm income, as farm programs have became central vis-a-vis the crisis of over-production and the financial crisis that have affected large segments of the American farming sector in the eighties.

The relationship between the State and the creation of off-farm jobs in marginal rural areas is twofold. First, given the reduction in farming revenues and the relative underdevelopment of these regions, the State is forced to attempt to increase the level of local accumulation. The accomplishment of this goal is attempted through capitalizing on the lower costs of production available in these regions to invite corporations to invest there. In return the State guarantees discounts in the process of construction of the plant and the development of infrastructures. The result, according to the desires of the officialdom of the State, would be that of supplying a number of jobs to curb local unemployment and stimulate local demand so that the total process of accumulation would be reestablished. Second, at the level of legitimation, the creation of new jobs enhances the ability of State officials[36] to maintain support among their constituents and ensures social stability in the community.

In the process of attracting new jobs to the region the State is willing to settle for lower paying and often part-time, temporary employment to replace higher paying jobs lost in the past or simply to curb unemployment. However, this process creates contradictions at the levels of both accumulation and legitimation. As far as the former is concerned, low paying, part time jobs have traditionally reduced the capacity to generate external economies and, consequently, to trigger further socio-economic development. Given the non-local nature of these corporations, the incomes of local workers are often the only source available to achieve this goal (Summers, 1986). At the legitimative level, the expectations of unemployed workers may not be met by these kinds of new jobs and dissatisfaction among local residents is thus generated.

The State is also forced to maintain farm programs and support for financial institutions involved in farming. The reason for this State posture can be summarized as follows. Attempts to reduce or eliminate farm programs and to reduce availability of credit are resisted at various levels. This resistance comes primarily at the farming community level as farmers depend increasingly on farm programs and credit for the continuous existence of their farms, even in the case of small farms. Such attempts are also resisted by State officials who are appointed or work in farm dependent areas. This is the case, for instance, of senators and representatives of farming communities who work for an increase in the magnitude of funds in farm legislation, against the reduction of tobacco support programs and for the expansion of farm credit. Additional resistance to the elimination of farm programs is found among the general public of the United States, as farming is identified, even among non-rural residents, as a fundamental trait of American society. The echo of the farm aid campaign and the emotional mobilization that it has generated are cases in point. Consequently, the failure to maintain farm programs could easily trigger a legitimation crisis that would erode the very support that hegemonic groups enjoy in the U.S. today. The hegemonic project in this historical conjuncture has among its fundamental characteristics the appeal to traditional fundamental values, of which farming is among the most relevant.

Maintenance of farm programs, however, tends to worsen the fiscal crisis of the State. The magnitude of agricultural overproduction casts a strain on the State's ability to curb lower farm revenues. Moreover, major farm programs, such as farm income programs, have been, historically, incentives for increasing farm output among producers. Similarly, national priorities in research still emphasize production and productivity as principal focuses for the immediate future (see FY 1987 Priorities for Research, Extension and Higher Education. Joint Council on Food and Agricultural Sciences, 1985), furthering the need for State financial assistance. Once again, the contradiction between legitimation and fiscal crisis of the State observed for Italy is reproduced in the case of the United States.

If this scenario is correct, what are the possible consequences in marginal regions vis-a-vis the problems discussed above? Among the several hypotheses, two are relevant for this discussion.

First, unlike the case of Italy, it can be hypothesized that the processes of industrial decentralization and informalization are sufficient to sustain local development. Consequently, a network of industrial and service activities will be established so that stable jobs will now be available for rural residents. Small farms will thus lose their function as keeper of surplus labor to become suppliers of labor to the expanding non-farm sector. The economic function of the small farm sector will be exhausted in the short run, leaving the farm as a form of lifestyle. The persistence of small farms in marginal areas of the US will thus be linked largely to the choice of lifestyle by rural residents. As this attitude is strongly present today among rural residents, it seems fair to hypothesize that it will remain as such at least in the near future.

Second and in accordance with the case of Italy, the nature of the decentralization and informalization processes taking place in marginal rural areas does not allow for the establishment of traditional forms of labor/capital relations. The conditions for the existence of industrial activities thus remain those of a cheap means of production and State assistance. Once these conditions are violated and there is an increase in the costs of labor and/or other costs of production, the corporations involved either move their plants elsewhere or judge them unprofitable and dispose of them through selling or closing. In all of these cases the basic assumption is that economic growth is possible only as long as the conditions typical of regional underdevelopment are maintained. As in the case of Italy, growth with underdevelopment becomes the major characteristic of the region. The State is confronted with contradictory tasks and it is unable to provide immediate solutions. Its major goals of reproducing accumulation and maintaining legitimation are, in fact, dependent on the very processes of decentralization and informalization that are at the origin of the problem itself.

The persistence of small farms is, in this case, related, on one hand, to their role as supplier of marginal labor for decentralized and informal activities. On the other, these farms represent an element of security for workers vis-a-vis the precarious nature of off-farm employment.

In both the first and second hypothesis the persistence of small farms is based on conditions that do not involve only the farm as a productive unit. Rather, factors internal and external to the farm and the farm family operate to reproduce this situation.

As a final note, it is important to mention that the dialectical assumptions under which this study has been carried out do not allow for determination of future conditions, but rather for potentiality. This means that given the contradictory nature of historical events, their outcome can be understood only within the range of historically possible changes which are created by past and present human action and constitute the framework for the future. In this respect, though the persistence of small farms in the years to come is derived from the existing conditions in American society, a crystal ball of future farm structural changes is not offered. Such an attempt would be epistemologically reductive, as it limits the historical potentiality of human action to a reified motion that narrows the understanding of society to a preordained structure. Persistence is, then, the synthesis of a set of contradictory dynamic forces whose shape and strength are historically bounded.

Notes

1. In the United States the process of colonization provided a plentiful supply of new land, yet the increase of productivity of labor and land was still a major concern (Hambridge, 1978:12).

2. Among these works are: E. Flour De St. Genis, 1902, La Propriete' Rurale en France. Paris: Etienne. L. Jebb, 1907 The Small Holdings in England. London: Wislon Publisher. Wiliam Clark, 1948, Farms and Farmers. Boston: Page Publisher.

3. Here the term "capitalist" means a technically and economically advanced unit.

4. For a further explanation of this theory see the following review of Mottura and Pugliese's works.

5. The same concept is defined by Mottura and Pugliese as follows: "A situation in agriculture in which there are two distinct sectors involving different forms of the organization of production, different levels of productivity, different perspectives on consolidation, and finally different roles in relation to the overall process of develop-ment" (1980:171).

6. This situation also generated a lack of supply of agricultural products that needs to be compensated with imports. West Germany produces, in fact, only 50 percent of its internal consumption.

7. Agriculture was the location of the vast majority of the industrial reserve army (in this case in its latent form). Keeping this labor in place was not only relevant to controling the industrial labor demand, but it was also a means for not exacerbating the contradictions in urban areas.

8. The new drastic reduction of rural unemployment in the South and the increased industrial employment in the North were so impressive that several social scientists theorized the near end of the socio-economic inequalities between North and South in the not so distant future (Eckaus, 1965; Hildebrand, 1965; Lutz, 1962).

9. In all of the following cases questions about farm and off-farm income were asked. In order to obtain accurate answers some specific measures were taken. First, the interviewer stated the absolute anonymity of the interview. Second, it was stressed that the respondent was not obliged to answer any specific question. Finally, no exact income figures were asked, only ranges. Such measures insured a high rate of success, for almost all of the farm families interviewed provided answers to the income questions. The

ones who refused provided enough information so that the family income was easily estimable.

10. The Italian Census Bureau measures hired labor according to the number of work days paid by the farm entrepreneur. It follows that if two farmworkers are hired for one day each, the amount of labor employed will be equal to two days. In this study this type of measurements is employed.

11. The Italian welfare systems not only subsidizes poor individuals and families but also provides payments to workers who are either unemployed or disabled in some manner. A large portion of rural workers, estimated at 65 percent (Pugliese, 1983), receive at least one of the various forms of welfare payments available through the State. Once acquired, these benefits are rarely revoked. One example is provided by the unemployment pension for farmworkers. Periodically, a list of all the unemployed farmworkers in the country is compiled. All of these workers are entitled to receive unemployment benefits in accordance with the number of days worked during the previous year. Such payment is continued until unemployment ceases, and it is still maintained if the individual finds employment outside of agriculture. As Mingione (1981) suggests, these individuals receive a tenured pension.

12. The housing market in Italy is characterized by a shortage of low- to-medium-priced houses, which in turn generates relatively high prices on the market. An average house in Italy is 20 to 25 percent more expensive in real terms than a comparable one in the US (Italy, 1983).

13. There is some compensation in kind for this work such as free food and wine. Some olive oil is also given to each of the helpers. It is a custom to reciprocate the favor, so members of this family perform work on relatives' farms when requested.

14. The word theory is employed here in the sense of "explanation" rather than in its alternate meaning of "paradigm." As specified in Chapter IV, the paradigm employed in this study is the dialectic.

15. As Giddens (1971) and Rossi (1982) have suggested, Weber's position is not altogher supportive of an autonomous, all-comprehensive set of values that, developing out of themselves, explains the behavior of farmers. On the contrary, "the ideas which guide the behaviour of agricultural workers, while they are not simply the expression of economic interests, do not spring from nothing" (Giddens, 1971:123). What Weber meant to counter here is the philosophical school that analyzes historical

development in terms of generating ideas. Con-sequently, he supported a position in which economic factors also account for the complete explanation of social phenomena. However, "to suppose that these views developed in Weber's though simply within the context of an encounter with Marxism would be to greatly over-simplify the intellectual milieu in which Weber wrote" (Giddens, 1971: 123).

16. The complexity of this topic is treated further in the next chapter. It is important to mention that the continuous maintenance of State intervention in agriculture is related specifically to the level of class struggle in society. In a country like Italy in which political opposition has been strong historically, the State cannot use more coercive forms of action for social control of small farmers. Consequently, the action of patronage toward farmers is the most praticable technique. In other countries where political opposition is not as strong, the State can resort in other less costly forms of social control.

17. The productive role is that of producing agricultural commodities to be sold on the market in order to gain a profit sufficient to support the family.

18. This means that families depending only on the farm enterprise for their income are recipients of welfare payments. That is, they are families who are either poor or have one or more unemployed or disabled members.

19. Of the families interviewed, 166 out of 199 (83.4 percent) obtain the majority of their income from off-farm sources. Moreover, 64 percent of all families receive 90 percent or more of their income from off-farm origin.

20. In Chapter XXIV of Capital Marx defined accumulation: "Hitherto we have investigated how surplus-value emanates from capital; we have now to see how capital arises from surplus-value. Employing surplus-value as capital - reconverting it into capital - is called accumulation" (1972:365). In a footnote on the same page, Marx also acknowledged the definition of accumulation employed by Malthus. The note reads: "Accumulation of capital; the employment of a portion of revenue as capital (Malthus, 'Definitions'); Conversion of revenues into capital (Malthus, 'Principles of Political Economy' : 319, London, 1836)."

21. Marx considered as petty bourgeoisie all craftsmen, i.e., people who own their own labor and their own means of production. The bourgeoisie is the class composed of owners of the means of production who use hired labor (the proletariat).

22. Two major schools, the Keynesian and the Marxian, have contributed to the rejection of the theory of a self-corrective capitalist system. Working within a theoretical framework that accepts the general postulates of capitalism, Keynes (1971) argued for State intervention to end economic crisis which, if left to its course, would result in a more serious and prolonged crisis. The Marxian assumption (Sweezy, 1942) argued for recurrent and unavoidable crisis within capitalism that ultimately cannot be corrected by State intervention but only through a radical change in the system itself.

23. Labor is the most important one.

24. This is not to say that overt use of force is not possible in capitalism but that this violence must always be justified in light of the principles of freedom and equality among men.

25. As Mingione put it: "On the one hand the State attempts to ensure constant economic growth and therefore the profitability of public and private activities. On the other hand it endeavors to reproduce a social consensus based on full employment and the elimination of the more notable inequalities of income. These objectives are contradictory because the conditions of the former, technological process and the concentration of capital and expertise, lead to a shrinking employment base and a widening inequality of income." (1981:10).

26. These were acts forbiding rural-urban migration.

27. Accordingly, a cheaper labor force is available on the market, a labor force that can be hired at lower wage levels due to the supplementary income provided by the farm.

28. In several analyses of electoral patterns in Italy (De Luca, 1983) the "Ruralization" and "Southernization" of the DC's vote has been shown. Though a general decline in the political power of this party has been recorded nationwide (which culminated with the 1984 European election in which the Communist Party, PCI, replaced DC as the largest political party in the country), its influence in rural Southern regions is still overwhelming. As an official of the PCI put it: "If the rural South was not part of Italy, the country would have had a Communist government since the middle seventies" (interview appeared in La Repubblica, June 29, 1983, No. 151).

29. Although a quantification of the farm families directly involved in various aspects of the process of legitimation within the small farm sector is extraneous to the epistemological posture of this study, for the sake of completeness a quantitative description of the phenomenon

follows. Considering the fact that families classified in all of the groups participate in the actions of legitimation described above, a total of 85 percent of them are directly involved in this phenomenon. Involved with "material forms" of legitimation are 20.1 percent of the Reserve Group, 12.6 percent of the Complementary Group, and 8.5 percent of the Retired Group. Those families involved in "ideological legiti-mation" are 38.7 percent in the Traditional group and 5 percent in the Residual group.

30. The economic system's incapability of generating new jobs to offset the labor supply has been a growing phenomenon not only in Italy but also in the rest of Western Europe. In the last ten years, in spite of a growth in the available labor force, the number of officially recorded jobs on the continent has remained practically unchanged. Studies indicate (Bagnasco, 1981; Gershuny, 1979) a growth of "informal" economic activities that absorb, in a highly precarious and at times illegal manner, portions of the abundant labor supply.

31. In this respect the EEC policy on the matter does not leave any doubt with respect to the unlikelihood of this alternative taking place (see Fabiani, 1979:237-248).

32. The theory that describes persistence as the ability to perform well on the market limits the angle of investigation exclusively to the farm enterprise as such, making it an element closed in a hypothetical market and severed from all the other segments of society both at the structural and superstructural level. As indicated in previous pages, the relationship between these farms and the rest of society is very complex, and by no means can it be reduced to a unidimensional form.

33. An example of this political posture is given by the strategy of the U.S. Administration since the early 1980s. In this period there has been a constant attempt to reduce State intervention in the economy, both in terms of the deregulation of economic activities and the relative reduction of welfare expenses.

34. The informalization process has been studied and interpreted from various angles. The most frequent interpretations found in the literature are the following:

a) Informalization as a process of creative social change. According to this reading of the phenomenon, the informalization of economic activities is generated by the diffusion of advanced technologies, which changes the production and distribution of services to the masses. Informalization is, then, a fundamental part of the so-called "self-service society" where, through the use of these technologies, the public can self-produce services once delegated to the formal society (Gershuny, 1983).

b) Informalization is a phase of the rebellion against the overwhelming power of monopoly capital. In this reading, informal activities are seen as alternatives to the dominant forms of social organization on the same line as presented by Schumacher in "Small is Beautiful" and in the ecologist and youth movement against mature capitalism (Illich, 1981: Henize and Olk, 1982).

c) Particularly with regard to the third world and/or underdeveloped regions, informalization has been connected with the process of super-exploitation of multinational corporation on local labor. More specifically, informalization is a means by which costs of production are lowered and the generation of surplus values increased (Portes and Walton, 1981).

d) Informalization is connected with industrial restructuring and the decentralization of operations traditionally assigned to the primary segment of the economy. These operations are reassigned to the secondary segment and involve the substitution of highly paid, regular labor with low paid, irregular labor composed of weak members of the labor market such as women, the aged, illegal immigrants and students.

35. See the CBS Broadcast "60 Minutes" of February 17, 1985.

36. It is important to remind the reader that the notion of State official does not refer to members of Local or State governments nor the Federal government. Rather, it refers both to officials in State, Local and Federal governments and agencies and to individuals who may act representing these institutions though they do not belong permanently to any of them. Furthermore, is relevant to note that the action of the State should not be confused with the action of particular individuals. Rather, it but involved the total interaction between State officials and the bureaucratic apparatus within which they operate.

Bibliography

Adorno, T.W.
1973 Negative Dialectic. New York: The Seabury Press.

Allen, F.
1974 Socio-cultural Dynamics. New York: MacMillan.

Amin, S.
1978 Accumulation on a World Scale. New York: Monthly Review Press.

Ardigo, A. and Donati P.
1976 (ed.) Famiglia e Industrializzazione. Milan: Feltrinelli.

Arkleton,
1983 "Part-Time Farming in the Rural Development of Industrialized Countries." Report of a Seminar held in Scotland from 16 to 21 October, 1983. The Arkleton Trust.

Bagnasco, A.
1981 "La Questione Dell' Economia Informale." Stato e Mercato 1(April):173-196.

Bandini, M.
1956 La Riforma Fondiaria. Rome: Edizione Cinque Lune.

Barbagli, M.
1977 Famiglia e Mutamento Sociale. Bologna: Il Mulino.

Barberis, C.
1970 Gli Operai Contadini. Bologna: Il Mulino.

Barberis, C. and V. Sesto
1974 Produzione Agricola e Strati Sociali. Milan: Franco Angeli.

Battistini, G.
1983 "Sfiorato il Sorpasso Berlinguer Torna in Gioco." La Repubblica, 151(June 29):4-5.

Beale, C.
1978 "People and Land." In T. R. Ford (ed.), Rural USA: Persistence and Change. Ames: Iowa State University Press.

Benito, S.
1984 "La Situacion de la Agricultura Familiar." In La Nueva Agricultura Espanola. Madrid: PDEE.

Bertolini, P. and B. Meloni
1979 "Elementi di Dualismo in Agricoltura: Il Caso Piemontese." In Grimaldi P. (ed.), Condizione Contadina. Turin: Stampatori.

Berry, W.
1978 The Unsettling of America: Culture and Agriculture. San Francisco: Sierra Club Books.

Birnie, A.
1961 An Economic History of Europe. London: Methven.

Block, F.
1977 "The Ruling Class Does Not Rule." Socialist Revolution 7:6-28.
1980 "Beyond Relative Autonomy: State Managers as Historical Subjects." In R. Miliband and J. Saville (eds.), Socialist Register London: Merlin Press.

Boccella, N. M.
1982 Il Mezzogiorno Sussidiato. Milan: Franco Angeli.

Bogart, E.
1942 Economic History of Europe 1760-1939. New York: Longmans Green.

Bolaffi, G. and A. Varotti
1973 Agricoltura Capitalista e Classi Sociali in Italia. Bari: De Donato.

Bonanno, A.
1984 Agricoltura e Sviluppo Dualistico. Milan: Franco Angeli.

Bonanno, A. and C. Ritter
1983 "Accumulation, Legitimation and Dualism in the Farm Sector: The Case of some Advanced Western Society." Paper Persented at the RSS Annual Meeting, Lexington, Ky.

Bonanno, A. and T. Calasanti
1986 "The Political Economy of Rural Elderly: The Case of Small Farmers In Southern Italy." Ageing and Society, 6 (March): 13-37.

Brewster, D. et al.
1983 (eds.), Farms in Transition: Interdisciplinary Perspectives on Farm Structure. Ames: Iowa State University Press.

Brodrick, G. C.
1881 English Land and English Landlords. London: Ford.

Brooks, N.
1985 Minifarms, Farm Business or Rural Residence. USDA, Agricultural Information Bulletin No. 480.

Brusco, S.
1979 Agricoltura Ricca e Classi Sociali. Milan: Feltrinelli.

Buttel, F. and W. Flinn
1975 "Sources and Consequences of Agrarian Values in American Society." Rural Sociology, 40:134-151.

Buttel, F. and H. Newby
1980 The Rural Sociology of the Advanced Societies. Montclair N. J.: Allanheld Osmun.

Cacciola, R.
1980 "Struttura della Popolazione ed Emigrazione in una Provincia Meridionale: Caltanissetta." Incontri Meridionali 1(January-March):25-37.

Calza-Bini, P.
1974 "Contadini, Proletari o Vasto Ceto Medio?." La Critica Sociologica 30(December):180-203.
1976 Economia Periferica e Classi Sociali. Napoli: Liguori.

Capecelatro, E. M. and A. Carlo
1972 Contro la questione Meridionale. Rome: Samona.

Carlin, T. and Crecink, J.
1979 "Small Farm Definition and Public Policy." American Journal of Agricultural Economics 61(December):157:169.

Carnoy, M.
1984 The State and Political Theory. Princeton N. J.: Princeton University Press.

Castells, M.
1980 The Economic Crisis and American Society. Princeton N. J.: Princeton University Press.

1984 "Toward the Informal City? High Technology, Economic Change, and Spatial Structure." Working Paper No. 430, Institute of Urban and Regional Development, University of California, Berkeley.

Castelucci, L. et al.
1984 "Famiglie e Aziende Contadine in un Area di Recente Industrializzazione." La Questione Agraria 3(June):49-160.

Castronovo, V.
1975 "La Storia Economica." In Storia D'Italia. Vol. 4.1 Turin: Einaudi.

Cavazzani, A.
1980 Il Part-Time agricolo. Venezia: Marsilio.

C. E. C.
1979 Report on the Regional Problems in the Enlarged Community. Brussels: Commission of the European Communities.

Chantfort, E. V.
1982 "Will Midsize Farms Fade as Small Farms and Big Farms Multiply." Farmline 3(Aprile):4-6.
1983 "Rural Jobs at Risk." Farmline 2(March): 9-10

Chessari, G.
1980 "L'Altra Sicilia: Agricoltura e Sviluppo a Ragusa." Incontri Meridionali 1(January-March):20-35.

Chombart De Lauwe, L.
1979 L'Aventure Agricole de la France de 1945 a' Nos Jours. Paris: PUF.

Clapham, J. H.
1932 An Economic History of Modern Britain. London: C.U.P.

Clark, W.
1948 Farms and Farmers. Boston: Page Publisher.

Clout, D. H.
1975 Regional Development in Western Europe. London: John Wiley and Son.

Clout, D. H.
1976 The Regional Problem in Western Europe. New York: Cambridge University Press.

Colman, G. and S. Elbert
1984 "Farming Families: 'The Farm Needs Everyone'." In Harry K. Schwarzweller (ed.), Research in Rural Sociology and Development. Greenwich, Connecticut: Jay Press.

Comte, A.
1970 Introduction to Positive Philosophy. Indianapolis: Bobbs Merril.

Coughenour, C.M.
1977 "Attitudes of Kentucky Farmers About Farming, Values, Aspirations, Goals and Management." Lexington: University of Kentucky, Agriucltural Experiment Station.

Coughenour, C.M. and A. Gabbard
1977 "Part-Time Farmers in Kentucky in the Early 1970s: The Development of Dual Careers." Lexington: University of Kentucky, Agricultural Experiment Station.

Coughenour, C. M. and L. Swanson
1983 "Work Statuses and Occupation of Man and Women in Farm Families and the Structure of Farms." Rural Sociology 48(Spring):78-86.

Crecink, J.
1979 "Families with Farm Incomes, Their Income, Income Distribution and Income Sources." Washington, D.C.: USDA.

Danbom, D.
1979 The Resisted Revolution. Ames: Iowa State Press.

Daneo, C.
1969 Agricoltura e Sviluppo Capitalistico in Italia. Turin: Einaudi.

De Benedictis, M.
1980 "Sviluppo e Ristagno Dell' Agricoltura Nel Mezzogiorno." In M. De Benedictis (ed.), L'Agricoltura Nello Sviluppo del Mezzogiorno. Bologna: Il Mulino.

De Janvry, A.
1980 "Social Differentiation in Agriculture and the Ideology of Neopopulism." In F. Buttel and H. Newby (ed.), The Rural Sociology of the Advanced societies, Allanheld, Osmun, Montclair, N. J.

Del Monte, A. and B. Giannola
1978 Il Mezzogiorno nell'Economia Italiana. Bologna: Il Mulino.

De Luca, F.
1983 "La DC e' Ormai il Partito della Provincia." La Repubblica 151(June 29):2-3.

Denton, G. R.
1969 (ed.) Economic Integration in Europe. London: Wiedenfeld and Nicholson.

De Rosa, L.
1973 La Rivoluzione Industriale in Italia e il Mezzogiorno. Bari: La Terza.

Devine, J.
1985 "State and State Expenditure: Determinants of Social Investments and Social Consumption Spending in the Postwar United States." American Sociological Review 2(April):150-165.

Diaz, Q. P., et al.
1983 "El mercado y Los Precios de la Tierra." In La Nueva Agricultura Espanola. Madrid: PDEE.

Dickens, D., and A. Bonanno
1983 "Theoretical Problems in the Analysis of Development and Underdevelopment." Paper presented at the MSS annual meeting, Kansas City, Missouri.

Dickson, D.
1979 The Politics of Alternative Technology. New York: Universe Books.

Dobb, M.
1947 Studies in the Development of Capitalism. New York: International Publishers.

Durkheim, E.
1964 The Division of Labor in Society. George Simpson (Translator) Glencoe Ill.: Free Press.

Eckaus, R. S.
1965 "The Factor-Proportion Problem in Underdeveloped Areas." The American Economic Review 3(Sept.):119-136.

Ernele, L.
1936 English Farming Past and Present. New York: Longmans.

Emmanuel, A.
1972 Unequal Exchange. New York: Monthly Review Press.

Fabiani, G.
1979 L'Agricoltura Italiana tra Sviluppo e Crisi. Bologna: Il Mulino.

Fabiani, G. and M. Gorgoni
1973 "Un'Analisi delle Strutture dell'Agricoltura Italiana." Rivista di Economia Agraria 6(2):78-89.

Fanfani, R.
1977 "Crisi e Ristrutturazione Dell'Agricoltura Italiana." Inchiesta 26(March-April):6-25).

Faux, G.
1973 "Reclaiming America." Working Papers 2(Summber):37-42.

Flinn, W. and D. Johnson
1974 "Agrarianism among Wisconsin Farmers." Rural Sociology 39(Summer):187-204.

Flour De St. Genis, E.
1902 La Propriete' Rurale en France. Paris: Etienne.

Frank, A. G.
1972 "Who is the Immediate Enemy?". In J. C. Cockcroft et al. (eds.), Dependence and Underdevelopment. Garden City, New York: Anchor.
1979 Dependent Accumulation and Underdevelopment. New York: Monthly Review Press.

Franklin, S. H.
1971 The European Peasantry. London: Methven.

Friedland, W., M. Furnari, and E. Pugliese
1980 "The Labor Process in Agriculture." Paper presented at the Working Conference on the Labor Process, University of California, Santa Cruz.

Friedmann, H.
1978 "World Market State and Family Farms: Social Bases of Household Production in the Era of Wage Labor." Comparative Studies in Sociology and History 20(4):544-586.

Fua, G.
1969 Lo Sviluppo Economico In Italia. Milan: Franco Angeli.

Fuguitt, G., A.M. Fuller, R. Gasson, and G. Jones
1977 "Part-Time Farming: Its Nature and Implications." Ashford, Kent, England: Wye College, Center for European Agricultural Studies. Seminar Paper Number 1.

Furnari, M.
1977 "Articolazione Sociale e Territoriale della Occupazione Agricola in Italia." Rivista di Economia Agraria 3 (September):119-136.

Furnari, M.
1980 "Evoluzione e Struttura dell'Occupazione Agricola nelle Regioni Meridionali." In M. De Benedictis (ed.), L'Agricoltura nell Sviluppo del Mezzogiorno. Bologna: Il Mulino.

Garcia De Blas, A.
1983 "Empleo y Rentas en el Sector Agrario." In La Nueva Agricultura Espanola. Madrid: PDEE.

G A O
1978 Changing Character and Structure of American Agriculture. Washington, D.C.: GAO.
Gerschenkron, A.
1962 Economic Backwardness in Historical Perspective. Cambridge, Mass.: Belknapp Press of Harvard University Press.
Gershuny, J. I.
1979 "The Informal Economy, Its Role in a Post Industrial Society." Futures 1(2):92-108.
1983 The New Service Economy: The Transformation of Employment in Industrial Societies. New York: Preager.
Giddens, A.
1971 Capitalism and Modern Social Theory. Cambridge: Cambridge University Press, England.
Ginatempo, N.
1975 La Casa in Italia. Milan: Mazzotta.
Giori, D.
1983 Old People Public Expenditure and the System of Social Service: The Italian Case." In Anne-Marie Guillemard (ed.), Old Age and the Welfare State. Beverly Hills, Ca.: Sage Publications.
Gomes, M.
1983 "La Reforma de la Estructuras Agrarias." In La Nueva Agricultura Espanola. Madrid: PDEE.
Gorgoni, M.
1977 "Sviluppo Economico Progresso tecnologico e Dualismo nell' Agricoltura Italiana." Rivista di Economia Agraria 2(3):85-129.
Goss, K. and R. Rodefeld
1977 "Mechanization of US Agriculture From 1935 to 1974: Causes Direct and Indirect Effects." Paper Presented at the RSS annual meeting, Madison, Wi.
Goss, K., R. Rodefeld, and F. Buttel
1980 "The Political Economiy of Class Structure in US Agriculture: A Theoretical Outline." In F. Buttel and H. Newby (eds.), The Rural Sociology of Advanced Societies. Montclair, N. J.: Allanheld Osmun.
Gramsci, A.
1973 Prison Notebook. New York: International Publishers.
1974 La Questione Meridionale. Rome: Editori Riuniti.

Graziani, A.
1979 (ed.) L'Economia Italiana dal 1945 ad Oggi, II edition. Bologna: Il Mulino.
1979a "Il Mezzogiorno nel Quadro Dell'Economia Italiana." In A. Graziani, and E. Pugliese (eds.), Investimenti e Disoccupazione nel Mezzogiorno Bologna: Il Mulino.

Guaraldo, A.
1979 "Dalle Strutture Aziendali Agricole alla Formazione Sociale." In P. Grimaldi (ed.), Condizione Contadina. Turin: Stampatori.

Green, B.L. and T. Carlin
1985 Agricultural Policy, Rural Counties and Political Geography. USDA Report No. AGE850429.

Habermas, J.
1974 Theory and Practice. Boston: Beacon Press.
1975 Legitimation Crisis. Boston: Beacon Press.
1979 Communication and the Evolution of Society. Boston: Beacon Press.
1984 The Theory of Communicative Action, I: Reason and the Rationalization of Society. (Translated by T. Mc Carthy), Boston: Beacon Press.

Hambridge, D.
1978 "The Nature and Magnitude of Changes in Agricultural Technology." In Rodefeld et. al. (ed.), Changes in Rural America. St Louis: The Mosby Company.

Heffernan, W. and J. Heffernan
1985 "The Challenge for Farm Families: Can They Overcome?" Paper presented at the Show Me Extension Homemakers Meeting, Clinton, MO. September 30.

Heffernan, W. et al.
1982 "Small Farms: A Heterogeneous Category." The Rural Sociologist 2(March):62-71.

Heinze, R.G. and T. Olk
1982 "Development and Informal Economy." Future 14(June):189-204.

Hildebrand, G. H.
1965 Growth and Structure in the Economy of Modern Italy. Cambridge, Mass.: Harvard University Press.

Holloway, J. and S. Picciotto
1977 "Capital Crisis and the State." Capital and Class 2(March):64-80.

Hoogvelt, A.
1982 The Third World in Global Development. London: MacMillan Press.

Horan, P. and C. Tolbert
1984 The Organization of Work in Rural and Urban Labor Markets. Boulder, Co.: Westview Press.

Horkheimer, M.
1972 "Tradition and Critical Theory." In Critical Theory: Selected Essays. New York: Harper and Herder.

Illich, I.
1981 Shadow Works. Boston: M. Boyars.

ISTAT
1952 Censimento Generale Della Popolazione. Roma: ISTAT.
1980 Dati Sulle Caratteristiche Strutturali della Agricoltura. Rome: ISTAT.
1983 Dati Sulle caratteristiche Strutturali della Agricoltura. Rome: ISTAT.

Italy
1983 Documents and Notes. Rome: Department of Information.

Jebb, L.
1907 The Small Holdings of England. London: Wilson Publisher.

Jessop, B.
1983 Theories of State. New York: New York University Press.

Kautsky, K.
1971a La Questione Agraria (Die Agrarfrage). Milan: Feltrinelli.

Kautsky, K.
1971b Il Programma di Erfurt. Rome: Samona' e Savelli.

Keane, J.
1978 "The Legacy of Political Economy: Thinking With and Against Claus Office." Canadian Journal of Political and Sociological Theory 2(3):49-92.

Keynes, J. M.
1971 Collected Writings. New York: St. Martin's Press.

Kliebenstein, J., W. Heffernan, D. Barret, and C. Kirtley
1981 "Economic and Social Motivational Factors in Farming." Journal of the American Society of Farm Managers and Rural Appraisers 1(April):10-14.

Klatzmann, J.
1978 L'Agriculture Francaise. Paris: Edition Du Sevil.

La Repubblica
1983 Edizione Post-Elezioni Politiche. Roma: 29 Giugno.

Levy, H.
1911 Large and Small Holdings. London: Cambridge University Press.

Lin, W., G. Coffman, and J. Penn
1980 "US Farm Numbers, Size and Related Structural Dimentions." Washington D.C.: USDA Tecnichal Bulletin No. 1625.

Lukacs, G.
1971 History and Class Consciousness. Cambridge: The MIT Press.

Lutz, V.
1962 Italy, A Study in Economic Development. London: Oxford University Press.

Mapp, H.P., M. Hardin, O. Walker and T. Persand
1979 "Analysis of Risk Management Strategies for Agricultural Producers." American Journal of Agricultural Economics 5(December): 1071-77.

Mann, S., and J. Dickinson
1980 "State and Agriculture in Two Eras of American Capitalism." In F. Buttel and H. Newby (eds.), The Rural Sociology of the Advanced Societies. Montclair N. J.: Allanheld Osmun.

Manscholt, S.
1968 Agricultural Plan for the '80s. Brussels: European Economic Community.

Marcuse, H.
1968 Negations: Essay in Critical Theory. Boston: Beacon Press.

Martinez, D.
1985 "The Bottom Has Faded, and Many Rural Areas Lag Behind." Farmline 2(April):12-14.

Marx, K.
1965 Basic Writings on Politics and Philosophy. Edited by A. J. Jones, New York: McMillan.
1972 Capital. Moscow: Progress Publishers.
1973 Grundrisse. New York: Vintage Book. Lewis S. Fever. Garden City, New York: Anchor Books.

Mayhew, A.
1970 "Structural Reform and the Future of West Germany Agriculture." Geographical Review 49(1):54-68.

1971 "Agrarian Reform in West Germany." Transactions of the Institute of British Geography 52(3):37-49.

Merleau-Ponty, M.
1973 Adventures of the Dialectic. Evanston Ill.: Northwestern University Press.

Mingione, E.
1981 (ed.) Classi Sociali e Agricoltura Meridionale. Milan: Giuffre'.
1981a Social Conflict and the City, New York: St. Martin's Press.

Mottura, G.
1975 "Sviluppo Capitalistico nelle Campagne dal Dopoguerra ad oggi." Citta' Classe 3(July):18-31.

Mottura, G. and E. Pugliese
1975 Agricoltura Mezzogiorno e Mercato del Lavoro Bologna: Il Mulino.

Mottura, G. and E. Pugliese
1980 "Capitalism in Agriculture and Capitalist Agriculture: The Italian Case." in F. Buttel and H. Newby (eds.), The Rural Sociology of the Advanced Societies. Montclair N. J.: Allanheld Osmun.

Munoz, R.
1984 "Socioeconomic Characteristics of Small Family Farms in Mississippi and Tennessee." The Rural Sociologist 1(January):3-7.

Mutti, A., and I. Poli
1975 Sottosviluppo e Meridione. Milan: Mazzotta.

Nikolitch, R.
1972 Family-Size Farms. Washington, D.C.: USDA, Economic Research Service, Agricultural Economic Report #449.

Noble, D.
1977 America By Design. New York: Oxford University Press.

Nocifora, E.
1980 "Sviluppo Economico Siciliano e Squilibri Territoriali." Incontri Meridionali 1(January-March):70-93.

O'Connor, J.
1973 The Fiscal Crisis of the State. New York, St. Martin's Press.

Offe, C.
1972 Structural Problems of the Capitalist State. London: McMillan.

1973 "The Capitalist State and the Problem of Policy Formulation." In L. N. Linberg et al. (ed.), Stress and Contradictions in Modern Capitalism. Lexington, Mass.: D. C. Heath.

1984 Contradictions of the Welfare State. John Keane (ed.) Cambridge, Mass: MIT Press.

Offe, C. and V. Ronge

1979 "Theses on the Theory of the State." In J. W. Freiberg (ed.), Critical Sociology. New York: Irvington Publishers.

Orloff, A.S. and T. Skocpol

1984 "Explaining the Politics of Public Social Spending." American Sociological Review 49(December):726-750.

Paci, M.

1973 Mercato del lavoro e Classi Sociali in Italia. Bologna: Il Mulino.

Perez Blanco, J. M.

1983 "Rasgos Macroeconomicos Basicos de la Evolucion de la Agricultura Espanola 1964-82." In La Nueva Agricultura Espanola. Madrid: PDEE.

Perez Diaz, V.

1983 "Los Nuevos Agricultores." In La Nueva Agricultura Espanola. Madrid: PDEE.

Perez-Saiz, S. P.

1981 "Capitalist State and Fetishization." In P. Zarembka (ed.), Research in Political Economy, Greenahich, Con.: Jay Press.

Pezzino, P.

1972 "The Agrarian Reform in Italy." Monthly Review (Italian Edition) 8-9(2-3):158-174;79-103.

Pfeffer, M.

1983 "Industrial Farming." Democracy 3(Spring) :37-49.

Phall, R. and J. I. Gershuny

1982 "Unemployment and Domestic Division of Labor." Paper Presented at the X World Congress of Sociology, Mexico City.

Piccone, P.

1983 Italian Marxism. Berkeley: University of California Press.

Podblieski, C.

1974 Italy: Development and Crisis in the Post-War Economy. Oxford: Clarendon Press.

Portes, A.
1976 "On the Sociology of National Development: Theories and Issues." American Journal of Sociology 82(July):55-85.

Portes, A. and A. Stepick
1985 "Unwelcome Immigrants: The Labor Market Experience of 1980 (Mariel) Cuban and Haitian Refugees in South Florida." American Sociological Review 4(August):493-514.

Portes, A. and J. Walton
1981 Labor, Class, and the International System. New York: Academic Press.

Poulantzas, N.
1978 State Power Socialism. London: NLB.

Procacci, G.
1971 Introduction to the Italian Edition of The Erfurt Program. Milan: Feltrinelli.

Progetto, Sicilia
1978 Dati e Caratteristiche Dell' Economia Siciliana. Palermo, Sicily.

Pugliese, E.
1977 Note sulla Situazione Agricola italiana. Lecture presented at the University of Messina (Italy).
1983 I Braccianti Agricoli in Italia. Milan: Franco Angeli.

Pugliese, E. and M. Rossi
1975 "Dualismo Strutturale in Agriculture e Mercato del Lavoro." In A. Graziani (ed.), Crisi e Ristruturazione dell' Economia Italiana. Turin: Einaudi.

Pye, L. and S. Verba
1965 Political Culture and Political Development. Princeton N. J.: Princeton University Press.

Raup, P.
1978 "Some Questions of Value and Scale in American Agriculture." American Journal of Agricultural Economics 2(May):303-308.

Reimund, P.
1982 Interview Reported in Chantford, "Will Midsize Farms Fade as Small and Big Farms Multiply?." Farmline 3(April):4-6.

Riley, M. et al.
1972 Agin and Society, Vol. 3; A Sociology of Age Stratification. New York: Russel Sage Fundation.

Rodefeld, R. at all
1978 (eds.) Change in Rural America. St. Louis: the Mosby Company.

Rodgers, A. B.
1970 "Migration and Industrial Development: The South Italian Experience." Economic Geography 46(3):43-59.

Ross, P., H. Bluestone, and K. Hines
1978 Indicators of Social Well-Being for US Counties. Washington, D. C.: USDA.

Rossi, P.
1982 Max Weber, Razionalita' e Razionalizzazione. Milan: Il Saggiatore.

Rossi-Doria, M.
1979 "Breve Storia dei Contadini italiana dallo Inizio del Secolo and Oggi." Inchiesta 38-39(March-June):87-104.

Russi, A.
1971 "Analisi delle Statistiche Capitalistiche: I Censimenti dell'Agricoltura." Inchiesta 1(January):27-44.

Saccomandi, V.
1978 "Politica delle Strutture e Riforma della P.C.A." Rivista di Economia Agraria 2(March):126-145.

Salvati, M.
1973 "Lo Scambio Ineguale: Un Intervento Polemico." In AA.VV., Salari, Sottosviluppo, Imperialismo. Turin: Einaudi.

Saraceno, C.
1976 Anatomia della Famiglia. Bari: Laterza.

Sassen-Koob, S.
1984 The New Labor Demand in Global Cities. In Michael P. Smith (ed.) Cities in Transformation. Beverly Hills: Sage Publication.

Say, J. B.
1834 A Tratise on Political Economy. Philadelphia: Criland Elliot.

Schroeder, E., F.C. Fligel, and J.C. Van Es
1985 "Measurement of the Lifestyle Dimensions of Farming for Small Scale Farmers." Rural Sociology 3(Fall):305-322.

Schroyer, T.
1970 "Toward a Critical Theory for Advanced Industrial Society." In Hans Peter Crietzel (ed.), Recent Sociology, No. 2, New York: McMillan.

Sereni, E.
1968 Il Capitalismo Nelle Campagne. Turin: Einaudi.

Sgritta, G., and A. Saporiti
1980 Family, Labor Market and the State in Italy From 1945 to the Present. Rome: University of Rome.

Sherry, R.
1976 "Comments on O'Connor's Review of the Twisted Dream." Monthly Review 28(May):52-60.

Smith, A.
1976 an Inquiry into the Nature and Cause of the Wealth of Nations. Oxford: Claredon Press.

Smith, D. and D. Capstick
1976 "Establishing Priorities Among Multiple Management Goals." Southern Journal of Agricultural Economics 2(December):37-43.

Spencer, H.
1967 The Evolution of Society. Selected from H. Spencer's Principles of Sociology. Chicago: University of Chicago Press.

Stefanelli, R.
1968 Agricoltura e Sviluppo Economico. Rome: Edizioni Sindacali.

Summers, G.F.
1986 "Rural Industrialization." The Rural Sociologist. Vol. 3(May): 181-187.

Swanson, L. and A. Bonanno
1985 Farm Crisis: A Dialectical Analysis. Unpublished Manuscript. Lexington: University of Kentucky.

Swanson, L. and J. Skees
1985 Public Policy for Farm Structure and Rural Well-being in the South. OTA Report. Washington, D.C.

Sweezy, P.
1942 The Theory of Capitalist Development. New York: Monthly Review Press.

Taylor, J.
1979 From Modernization to Modes of Production. London: The MacMillan Press LTD.

Toennies, F.
1957 Community and Society. New York: Harper and Row.

Tracy, M.
1964 Agriculture in Western Europe. New York: Preager.

USDA
1978 Census of U. S. Agriculture. Washington, D. C.: USDA.
1980 Economics and Statistics, Program Results and Plans. Washington, D. C.: USDA.
1984 Chartbook of Nonmetro-Metro Trends. USDA, Rural Development Research Report No. 43.
1984a Handbook of Agricultural Charts. USDA, Agricultural Handbook No. 637.
1985 The Current Financial Conditions of Farmers and Farm Lender. USDA, ERS, Agricultural Information Bulletin No. 490.

USDC
1985 Statistical Abstract of the United States, 1985. US Department of Commerce.

Velde, P.
1983 "Profile of the Super Farm." Farmline 7 (August-September) :4-7.

Villari, R.
1966 Il Sud Nella Storia D'Italia. Bari: La Terza.

Vogeler, I.
1980 The Myth of the Family Farm, Boulder, Colorado: Westview Press.

Wallerstein, I.
1974 The Modern World System. New York: Academic Press.

Weber, M.
1949 the Methodology of the social Sciences. Glencoe, Ill.: Free Press.

Weber, M.
1958 From Max Weber: Essay in Sociology, Capitalism and Rural Society in Germany. H. H. Gerth and C. Wright Mills (eds.), New York: Oxford University Press.

Weber, M.
1978 Selections in Translation. W. C. Runciman (ed.), New York: Cambridge University Press.

White, S.
1983 "Return Migration to Appalachian Kentucky: An Atypical Case of Nonmetropolitan Migration Reversal." Rural Sociology 3(Fall):471-491.

Wolfe, A.
1977 The Limits of Legitimacy, Political Contradictions of Late Capitalism. New York: Free Press.

Young, A.
1809 General View of the Agriculture of the County of Oxfordshire. London. No Publisher.

Zeller, A.
1970 L'imbroglio Agricole du Marche' Commun. Paris: Calmann-Levy.

Tables and Figures

Table 1
Agriculture as (a) a Proportion of Labor Force and (b) a Source of Gross National Product 1971 (%) in Selected Western Countries

	(a)	(b)		(a)	(b)
Belgium	4.4	4.2	Ireland	24.9	16.4
France*	9.0	5.0	United Kingdom	2.7	2.9
Italy**	15.0	9.0	Austria	17.9	6.0
Luxemburg	10.1	4.4	Norway	13.8	5.3
Netherlands	6.9	5.7	Portugal	31.1	16.2
West Germany	8.3	3.5	Spain***	15.1	5.9
USA***	3.1	3.5	Sweden	8.0	NA

*1978 figures
**1975 figures
***1982 figures

Sources: Euostat, Statistics of European Agriculture, CEE, 1978. US Census of Agriculture, 1982. La Nueva Agricultura Espanola, 1983.

Table 2
Number of Farms According to Size in Selected Western European Countries (percentages 1970 and 1975)

	1970		1975	
	Less Than 10 ha	More Than 50 ha	Less Than 10 ha	More Than 50 ha
West Germany	59.0	1.8	54.2	2.9
France	40.5	8.5	35.5	11.7
Italy	86.2	1.7	85.8	5.6
Holland	49.8	1.5	46.3	2.2
Belgium	59.1	2.0	52.1	3.2
Luxemburg	35.6	4.3	31.4	9.3
Great Britain	31.1	27.3	27.0	30.5
Ireland	38.0	7.5	NA	NA
Denmark	32.8	6.1	31.2	7.8

Source: Eurostat, Statistics of European Agriculture, CEE, 1978.

Table 3
Farms According to Size and Utilized Land as a Percentage of Total Utilized Land in the Country, in Selected Western European Countries (1970 and 1975)

	1970		1975	
	Less Than 10 ha	More Than 50 ha	Less Than 10 ha	More Than 50 ha
West Germany	21.5	12.3	16.8	16.7
France	9.1	35.5	6.8	41.8
Italy	38.0	32.6	37.0	47.9
Holland	19.0	9.3	15.5	12.3
Belgium	23.7	13.1	17.3	17.4
Luxemburg	8.3	14.3	6.0	27.4
Great Britain	2.5	79.0	2.1	8.8
Ireland	9.8	31.7	NA	NA
Denmark	9.1	25.5	7.7	30.2

Source: Eurostat, Statistics of European Agriculture, CEE, 1978.

Table 4
Size of Farms

Farm Categories	Number of Farms	Percent	Cumulative Percent
2 ha. or less	60	30.2	30.2
From 2.1 to 5 ha.	100	50.8	8.09
From 5.1 to 10 ha.	29	14.6	95.5
Over 10 ha.	9	4.5	100.0

Table 5
Size of Small Farms According to Utilized Land (SAU)

Farm Categories	Number of Farms	Percent	Cumulative Percent
2 ha. or less	83	41.7	41.7
From 2.1 to 5 ha.	104	52.3	94.0
Over 5 ha.	12	6.0	100.0

Table 6
Types of Crops Raised on Small Farms

Type of Cultures	Number of Farms	Percent	Cumulative Percent
Vegetables and fruits only	98	49.3	49.3
Vegetables, fruits and other cultures	78	39.0	88.3
Others	23	11.7	100.0

Table 7
Number of Parcels in Farms

Number of Parcels	Number of Farms	Percent	Cumulative Percent
1	108	54.3	54.3
2	34	17.1	71.4
3	24	12.1	83.5
4 or more	33	16.5	100.0

Table 8
Average Distance of Parcels

Distance to Parcels (Km)	Number of Farms	Percent	Cumulative Percent
1	92	46.3	46.3
2	33	16.6	62.9
3	34	17.1	80.0
4 or more	40	20.0	100.0

Table 9
Gross Farm Income

Farm Income in Millions lira	Number of Farms	Percent	Cumulative Percent
From 0 to 1.5	104	52.3	52.3
From 1.6 to 8	35	17.6	69.9
From 8.1 to 12	27	13.6	83.5
From 12.1 to 20	33	16.5	100.0

Table 10
Net Farm Income

Farm Income in Millions lira	Number of Farms	Percent	Cumulative Percent
From 0 to 1.5	82	41.2	41.2
From 1.6 to 6	74	47.2	78.4
From 6.1 to 10	27	13.6	92.5
Over 10	15	7.5	100.0

Table 11
Family Farm Income of Off-Farm Origin

Percentage of Total Income	Number of Farms	Percent	Cumulative Percent
20 or less	24	12.1	12.1
From 20.1 to 40%	9	4.5	16.6
From 40.1 to 60%	10	5.0	21.6
From 60.1 to 80%	18	14.1	35.7
Over 80%	128	64.3	100.0

Table 12
Farm Family Total Annual Income (lira)

Categories of Income (00000)	Number of Farms	Percent	Cumulative Percent
10 or less	43	21.6	21.6
From 10.1 to 20	65	32.7	54.3
From 20.1 to 28	78	39.2	93.5
Over 28	13	6.5	100.0

Table 13
Perception of Adequacy of Farm Income

Perception of Adequacy of Farm Income	Number of Families	Percent	Cumulative Percent
Sufficient	33	16.6	16.8
Partially sufficient	9	9.6	26.4
Insufficient	147	73.8	100.0

Table 14
Types of Off-Farm Employment in Farm Families

Types of Employment	Families	Percent	Cumulative Percent
More than one member full-time	28	14.0	14.0
One member full-time	89	44.8	58.8
Only part-time	34	17.1	75.9
No off-farm jobs	48	24.1	100.0

Table 15
Types of Off-Farm Jobs of Farm Family Members

Types of Jobs	Number of Workers	Percent	Cumulative Percent
Blue Collar Industrial Sector	49	20.8	20.8
Blue Collar Tertiary Sector	26	11.0	31.8
Agricultural	26	110.0	42.8
Clerical	115	48.7	91.5
Professional or Managerial	20	8.5	100.0
Total	236	100.0	100.0

Table 16
Hours of Farm Work Performed Weekly by Farm Family Members (except heads of households)

Hours	Family Members	Percent	Cumulative Percent
20 or less	166	30.1	30.1
From 21 to 49	79	14.3	44.4
Over 50	306	55.6	100.0

Table 16a
Hours of Farm Work Performed Weekly by Heads of Households

Hours	Farmers	Percent	Cumulative Percent
20 or less	19	9.5	9.5
From 21 to 49	93	46.7	56.3
Over 50	87	43.7	100.0

Table 17
Amount of Daily Wages Paid to Hired Labor

Amount of Wages (000 lira)	Number of Workers	Percent	Cumulative Percent
20 or less	16	15.4	15.4
From 21 to 39	78	75.0	80.4
40 or above	10	9.6	100.0

Table 18
Machinery Present on Small Farms

Types of Machines	Number of Farms	Percent	Cumulative Percent
No machines	43	21.6	21.6
One tractor only	88	44.2	65.8
Two tractors	25	12.6	78.4
More than two tractors and other equipment	43	21.6	100.0

Table 19
Farm Products Consumed on the Farm (% of total production)

Percent of Farm Production	Number of Farms	Percent	Cumulative Percent
20% or less	72	36.2	36.2
From 21 to 40%	30	15.2	51.3
From 41 to 60%	14	7.0	58.3
From 61 to 80%	9	4.5	62.8
Over 80%	74	37.2	100.0

Table 20
Age Distribution of Agricultural Population of Working Age in Calabria, Sicily and Region Surveyed (%)

Age	Calabria	Sicily	Survey
14 - 19	7.4	5.3	13.9
20 - 39	35.8	36.6	35.0
40 - 59	45.4	48.9	34.4
60 - 64	7.0	6.3	9.0
65 and over	4.4	2.9	7.7

Source: ISTAT, Censimento Della Popolazione, Roma, 1982.

Table 21
Farm Life Satisfaction of Heads of Households

Satisfaction	Number of Farmers	Percent	Cumulative Percent
Very satisfied	117	59.2	59.2
Satisfied	26	13.0	72.2
Dissatisfied	55	27.8	100.0

Table 21a
Farm Life Satisfaction of Families

Satisfaction	Families	Percent	Cumulative Percent
Very satisfied	77	39.7	39.7
Satisfied	26	13.4	53.1
Dissatisfied	91	46.9	100.0

Table 22
Heads of Household's Attitudes Toward Farm Income and Farm Life Satisfaction

	Sufficient Farm Income	Insufficient Farm Income	Total
Satisfied with farm life	41	102	143
	78.8%	70.3%	
	28.7%	71.3%	
Dissatisfied with farm life	11	43	54
	21.2%	29.7	
	20.4%	79.6%	
Total	53	146	197

Table 23
Composition of Farm Families

Number of Individuals	Number of Families	Percent	Cumulative Percent
2 or less	35	17.6	17.6
3	38	19.1	36.7
4	58	29.1	65.8
5	39	19.6	85.4
6 or more	29	14.6	100.0

Table 24
Percentage of Off-Farm Income out of the Total Family Income by Group; G = Group

Groups	Percent	G 1	G 2	G 3	G 4	G 5	G 6
Traditional	58.7		*				
Reserve	41.6						
Complementary	74.8	*	*				
Retired	80.8	*	*				
Equity	65.9		*				
Residual	85.3	*	*		*		
Total	63.3						

Significance : .0000

*Denotes pairs of groups significantly different at the .05 level.

Table 25
Neighboring Families' Motivations for Abandoning Farming by Group (%)

	Economic Motivat. %	Dissatisfactory Lifesttyle
Traditional	87.5	12.5
Reserve	90.0	10.0
Complementary	100.0	0.0
Retired	100.0	0.0
Equity	100.0	0.0
Residual	96.4	3.6

Not Significant at .05

Table 26
Percentage of Families with Heads of Household Who Are Satisfied with Farm Life and the Farm by Group

Groups	Percent
Traditional	75.0
Reserve	75.0
Complementary	76.0
Retired	88.2
Equity	75.0
Residual	46.4
Total	72.2

Significance : .0303

Percentage of Families with Family Members Who Are Satisfied with Farm Life and the Farm by Group

Groups	Percent
Traditional	62.9
Reserve	52.2
Complementary	52.0
Retired	58.8
Equity	66.6
Residual	21.4
Total	21.4

Significance : .0076

Table 27
Size of Farms (Hectares) by Group

Groups	Average Size
Traditional	7.37
Reserve	6.64
Complementary	11.08
Retired	7.00
Equity	6.54
Residual	7.30
Total	7.74

Significance : .0272

Table 28
Total Family Income by Group (million lira); G = Group

Groups	Average Income	G (2)
Traditional	10.23	*
Reserve	7.82	
Complementary	10.96	*
Retired	9.52	*
Equity	10.58	*
Residual	9.85	*
All Groups	9.74	

Significance : .0031

*Denotes pairs of groups significantly different at the .05 level.

Table 29
Net Farm Income 1982 by Group (million lira); G = Group

Groups	Average Income	G 4	G 6
Traditional	5.68		*
Reserve	6.15	*	*
Complementary	5.76		*
Retired	5.05		
Equity	6.16	*	*
Residual	4.57		
Total	5.60		

Significance : .0004

*Denotes pairs of groups significantly different at the .05 level.

Table 30
Family Members' Off-Farm Activities by Group
A) Families With No Members Working Off-Farm

(Percent of Each Group) Groups	Percent	(Percent of Total Sample) Groups	Percent
Traditional	20.8	Traditional	33.3
Reserve	37.5	Reserve	33.3
Complementary	4.0	Complementary	2.3
Retired	47.1	Retired	17.7
Equity	8.3	Equity	2.3
Residual	17.9	Residual	11.1
		Total	100.0 (45)

Significance : .0000

(continued)

B) Families With Only One or More Offspring Working Off-Farm by Group

(Percent of Each Group) Groups	Percent	(Percent of Total Sample) Groups	Percent
Traditional	29.2	Traditional	39.6
Reserve	40.0	Reserve	30.2
Complementary	4.0	Complementary	1.8
Retired	52.9	Retired	17.0
Equity	33.4	Equity	7.6
Residual	7.1	Residual	3.8
		Total	100.0 (53)
			Significance : .0485

C) Families With Only One or Both Parents Working Off-Farm

(Percent of Each Group) Groups	Percent	(Percent of Total Sample) Groups	Percent
Traditional	38.9	Traditional	40.0
Reserve	17.5	Reserve	10.0
Complementary	52.0	Complementary	18.5
Retired	0.0	Retired	0.0
Equity	50.0	Equity	8.5
Residual	57.1	Residual	23.0
		Total	100.0 (70)
			Significance : .0000

(continued)

D) Families with One or Both Parents and One or More Offspring Working Off-Farm

(Percent of Each Group) Groups	Percent	(Percent of Total Sample) Groups	Percent
Traditional	11.1	Traditional	30.7
Reserve	5.0	Reserve	7.7
Complementary	40.0	Complementary	38.5
Retired	0.0	Retired	0.0
Equity	8.3	Equity	3.8
Residual	17.9	Residual	19.3
		Total	100.0 (26)

Significance : .0045

TABLE 31
Families with Members Who Have One or More Off-Farm Part-time Jobs by Group; Percent and Absolute Values

Groups	Percent	Number
Traditional	18.2	14
Reserve	25.0	10
Complementary	8.0	2
Retired	11.8	2
Equity	16.6	2
Residual	3.6	1

Significance : .0103

Table 32
Types of Farm Acquisition by Group, Percent and Absolute Values

	Farm Inherited		Farms Purchased		Farms Inherited and Purchased		Land Reform	
Groups	%	#	%	#	%	#	%	#
Traditional	75.0	54	12.5	9	9.7	7	2.8	2
Reserve	45.0	18	20.0	8	15.0	6	20.0	8
Complementary	64.0	16	16.0	4	16.0	4	4.0	1
Retired	52.9	9	47.1	8	0.0	0	0.0	0
Equity	8.3	1	83.3	10	8.3	1	0.0	0
Residual	82.1	23	10.7	3	7.1	2	0.0	0
Signif.:	.0000		.0000		N.S.		.0320	

Table 33
Average Farm Wage Paid to Hired Labor by Group (thousand lira per day)

Groups	Lira
Traditional	25.6
Reserve	25.2
Complementary	25.5
Retired	30.0
Equity	33.2
Residual	36.3

Significance : .0388

Table 34
Families' Attitudes Toward Farm-Income by Groups. (Percent of Families Whose Members Consider Farm Income Insufficient to Support the Family.)

Groups	Percent
Traditional	67
Reserve	41
Complementary	96
Retired	88
Equity	100
Residual	73

Significance : .0000

Table 35
Average Age of Heads of Household by Group

Groups	Average Age
Traditional	59.8
Reserve	63.6
Complementary	59.6
Retired	68.5
Equity	60.5
Residual	61.4

Significance : .0026

Table 36
Average Farm Size: Total Land Utilized (SAU), (Hectares)

Groups	Size
Traditional	4.00
Reserve	3.97
Complementary	3.72
Retired	3.23
Equity	4.25
Residual	3.28
Total Average	3.80

Not Significant at .05

Table 37
Average Percentage of Farm Products Consumed by the Family: Six Groups

Groups	Size
Traditional	44.1
Reserve	45.4
Complementary	49.4
Retired	40.3
Equity	42.0
Residual	32.4
Total Average	42.9

Significance : .0193

FIGURE A   REGION SURVEYED

Index